Springer Theses

Recognizing Outstanding Ph.D. Research

For further volumes:
http://www.springer.com/series/8790

Aims and Scope

The series "Springer Theses" brings together a selection of the very best Ph.D. theses from around the world and across the physical sciences. Nominated and endorsed by two recognized specialists, each published volume has been selected for its scientific excellence and the high impact of its contents for the pertinent field of research. For greater accessibility to non-specialists, the published versions include an extended introduction, as well as a foreword by the student's supervisor explaining the special relevance of the work for the field. As a whole, the series will provide a valuable resource both for newcomers to the research fields described, and for other scientists seeking detailed background information on special questions. Finally, it provides an accredited documentation of the valuable contributions made by today's younger generation of scientists.

Theses are accepted into the series by invited nomination only and must fulfill all of the following criteria

- They must be written in good English.
- The topic should fall within the confines of Chemistry, Physics, Earth Sciences and related interdisciplinary fields such as Materials, Nanoscience, Chemical Engineering, Complex Systems and Biophysics.
- The work reported in the thesis must represent a significant scientific advance.
- If the thesis includes previously published material, permission to reproduce this must be gained from the respective copyright holder.
- They must have been examined and passed during the 12 months prior to nomination.
- Each thesis should include a foreword by the supervisor outlining the significance of its content.
- The theses should have a clearly defined structure including an introduction accessible to scientists not expert in that particular field.

Matthew R. Foreman

Informational Limits in Optical Polarimetry and Vectorial Imaging

Doctoral Thesis accepted by
Imperial College London, UK

Author
Dr. Matthew R. Foreman
Imperial College London
UK

Supervisor
Prof. Peter Török
Imperial College London
UK

ISSN 2190-5053
ISBN 978-3-642-28527-1
DOI 10.1007/978-3-642-28528-8
Springer Heidelberg New York Dordrecht London

e-ISSN 2190-5061
e-ISBN 978-3-642-28528-8

Library of Congress Control Number: 2012933840

© Springer-Verlag Berlin Heidelberg 2012
This work is subject to copyright. All rights are reserved by the Publisher, whether the whole or part of the material is concerned, specifically the rights of translation, reprinting, reuse of illustrations, recitation, broadcasting, reproduction on microfilms or in any other physical way, and transmission or information storage and retrieval, electronic adaptation, computer software, or by similar or dissimilar methodology now known or hereafter developed. Exempted from this legal reservation are brief excerpts in connection with reviews or scholarly analysis or material supplied specifically for the purpose of being entered and executed on a computer system, for exclusive use by the purchaser of the work. Duplication of this publication or parts thereof is permitted only under the provisions of the Copyright Law of the Publisher's location, in its current version, and permission for use must always be obtained from Springer. Permissions for use may be obtained through RightsLink at the Copyright Clearance Center. Violations are liable to prosecution under the respective Copyright Law.
The use of general descriptive names, registered names, trademarks, service marks, etc. in this publication does not imply, even in the absence of a specific statement, that such names are exempt from the relevant protective laws and regulations and therefore free for general use.
While the advice and information in this book are believed to be true and accurate at the date of publication, neither the authors nor the editors nor the publisher can accept any legal responsibility for any errors or omissions that may be made. The publisher makes no warranty, express or implied, with respect to the material contained herein.

Printed on acid-free paper

Springer is part of Springer Science+Business Media (www.springer.com)

In loving memory of my family

Supervisor's Foreword

Optical imaging and microscopy are not the most recent developments in physics. Since the invention of the optical microscope by the Dutch spectacle maker Zacharias Janssen in 1595 and the extensive contributions to practical, high resolution microscopy by the German physicist Ernst Abbe, a whole plethora of imaging modalities have been introduced, including dark field, fluorescence, interference contrast and finally culminating in phase contrast imaging for which the Dutch physicist Fritz Zernike was awarded the Nobel Prize in 1953. It was Ernst Abbe who employed geometrical optics theory to optical design aiming to solve the manufacturing problems of Carl Zeiss whose ailing eponymous company was producing lenses and other optical components on a trial and error basis. Abbe also established the scalar diffraction theory of optical imaging that lead to the derivation of his famous equation for the resolution of the microscope that is prominently carved into his memorial in front of the Friedrich Schiller University in Jena, Germany. Whilst the scalar theory of diffraction has proven to allow significant advances in optical imaging to be made, it is easily recognised that any complete treatment of optical imaging, and optical systems in general, must consider the polarisation state of light. A compound lens, for example, is frequently assumed to leave the polarisation state of light unchanged. This assumption breaks down due to refraction and Fresnel transmission at each optical surface, which can introduce longitudinal field components, and preferentially transmit one field component respectively. In addition stress in bulk glass is also known to modify the transmission of the polarisation of the electric field through the optical components. Modelling and computation of the structure of the electromagnetic field in such polarisation sensitive systems can be mathematically difficult and at points intractable, a problem that is further exacerbated by partial coherence properties of many physical light sources. This thesis sets out, in part, to mitigate some of these computational difficulties, particularly by combining vectorial ray tracing and modal analysis. This not only turns out to ease the computational load, but also adds a high level of physical understanding of imaging by complex optical systems.

Biological and material samples can frequently exhibit polarisation sensitive transmission properties, such as optical birefringence, hence resulting in further polarisation changes to those introduced by the optical system itself. Modelling difficulties not withstanding, these polarisation changes furnish significant opportunities to learn about the samples themselves, a fact frequently exploited in crystallography, bioimaging and remote sensing. For example, the director profile of liquid crystals can be studied by polarised light microscope or the measurement of time dependent rotations in the direction of a fluorescent dipole via polarisation changes can reveal its mobility. The limits of such polarimetric imaging systems are continually being advanced as more accurate measurements are sought so it is important to know the boundaries of validity of such measurements. Commonly measurement limits arise from the corrupting effect of noise present in the system and on the measurement. Whilst this issue is well understood in conventional irradiance measurement systems, and quantified using metrics such as the signal to noise ratio, alternatives are required when considering polarisation since the nature of measurement space is different. By employing concepts from information theory and statistical estimation suitable quantification of these limits can however be found, in turn allowing correct data analysis, system optimisation and benchmarking in polarimetry.

Finally this thesis looks beyond the theory to demonstrate the application of polarisation to current real-world problems. Optical data storage, for example, has seen profound development over the last few decades and has spawned a multi-billion dollar industry. Storage demands are however continually growing as the information age flourishes. These storage needs can happily be addressed by using data structures on the surface of optical media which alter the polarisation state of scattered light strongly. Following the earlier themes, modelling and optimisation of this proposition is presented in this thesis. A further example considered in this work, is that of measurements of the orientation of single molecules. Such endeavours are currently pertinent to fundamental biological research in which the tracking of single molecules can elicit information on cell function and structure, vital to the understanding of many biological processes and medical conditions.

London, January 2012   Peter Török

Acknowledgments

Writing a thesis is rarely a trivial task and much time and energy must be expended to arrive at fruition. That said this cost is often paid by those around be it on an intellectual, emotional, financial or other altruistic level. My experience is by no means an exception.

First and foremost, I must wholeheartedly thank Peter Török, to whom I owe an enormous debt of gratitude. Whilst, on paper, Peter's role was to act as my supervisor, in reality he did so much more and we have developed a strong friendship. I have not only learnt much from him about optics, science, food, religion and life in general, but in those inevitable times of trouble Peter stood as a pillar of strength and support, without which I do not where I would be now. His unflinching encouragement, understanding and wise advice were indispensable. Peter, I honestly can not express the depth of my gratitude and feelings. Thank you!

Given the frequent and numerous visits to the Török residence, my thanks also go to Peter's wife, Janey. Even with the rather forsaken hours of my visits, Janey always made me feel welcome and comfortable. This kindness does not go unnoticed and is greatly appreciated.

I would also particularly like to acknowledge Sherif S. Sherif. Although my time spent with Sherif in Ottawa was short, he taught me many things. Sherif introduced me to estimation and information theory and was patient enough to explain many concepts with which I was unfamiliar and struggled. This is knowledge which I am likely to carry, and use, for years to come. It is unfortunate that our paths went separate ways, however many thanks Sherif for sharing your knowledge.

My colleagues at Imperial College have all contributed to the success of this thesis in some manner. Arthur van de Nes, for example, has the patience of a saint, and unluckily bore the brunt of many of my theoretical and coding frustrations. Arthur would however always find the time to talk often resulting in interesting scientific discussions. Thank you Arthur, and my apologies! Peter Munro and Carl Paterson were also always willing to share their knowledge and expertise. They both have my gratitude, as too does Kenny Weir. Although not directly related to this thesis, but instead more to my development as a scientist, Kenny's explanations and demonstrations made my time teaching in the MSc laboratory both rewarding and constructive. Jesus Rogel-Salazar was also kind enough to volunteer for the arduous task of proofreading this thesis and to take the time to introduce me to the concept of punctuation!

Although frequently seen as a chore and nuisance, the need for funding is ubiquitous in all aspects of life. Science is sadly no exception. Fortunately through the continued financial support of the EPSRC I have however been given the opportunity to do what I love. This is by no means a small contribution and certainly worthy of record here. Grants from the Royal Academy of Engineering, the Institute of Physics and the Imperial Trust have further allowed me to venture to exciting and interesting places, to meet and interact with the leading minds in the field and to expand my knowledge base. Such a foundation is invaluable.

During my PhD I have been lucky to meet and make some great friends. Carlos and Alex, for instance, have been there from the very beginning and are responsible for my addiction to climbing. Both were highly influential in helping me settle and appreciate the joys of London. Christoferos and Chucky, although perhaps unbeknownst to them, have always served as a source of quiet inspiration and encouragement to me, driving me to better things. Furthermore Pavel and Ren have faced much of my stress and blues, yet have continually given their support and understanding, for which I am particularly thankful and hope to one day repay. Finally many thanks to Rafa and Ayari for their kindness and friendship. Together they have all helped keep life fun!

My old Oxford flatmates, Matt, Alex and Chris, have all been great friends for many years now. I am ever grateful for their generosity and friendship throughout the time I have known them and eagerly look forward to what is still to come. Needless to say, Kia and Julia have also played a central role and life would certainly not be the same without them. I love you all.

My love and thanks also go to my parents, without whom I would not have been in the position to write this text. Never demanding, yet always encouraging, their contribution is beyond measure: they have made me the person that I am today. I hope they both know how greatly I value my relationship with them and how thankful I am to have grown up in such a rich and loving environment.

Finally, yet by no means least, I am compelled to mention Gina. At the tender age of four one might wonder how a sweet little girl could have earnt my gratitude. Nevertheless, there is something to be said for the bliss of youth and Gina could always be trusted to give a bright, happy smile and immediately bring some of that peace back to my world, especially when I needed it the most. From the bottom of my heart, thank you Gina.

Contents

1 Introduction 1
    1.1 Motivation and Aims 1
    1.2 Thesis Structure and Overview 3
    References 4

2 Fundamentals of Probability Theory 7
    2.1 Random Variables 7
        2.1.1 Specifying a Random Variable 7
        2.1.2 Functions of a Random Variable 11
        2.1.3 Multiple Random Variables 11
    2.2 Random Processes 16
        2.2.1 Specifying a Random Process 16
        2.2.2 Stationarity and Ergodicity 16
        2.2.3 Spectral Analysis of a Random Process 18
    2.3 Some Important Types of Statistics 19
        2.3.1 Poisson Statistics 20
        2.3.2 Gaussian Statistics 22
    2.4 Conclusions 23
    References 24

3 Information and Estimation Theory 27
    3.1 Introduction 27
    3.2 Maximum Likelihood Estimation 29
        3.2.1 Definition 29
        3.2.2 Properties of the Maximum Likelihood Estimator .. 31
    3.3 Cramér-Rao Lower Bound 35
        3.3.1 Functions of Parameters and Nuisance Parameters . 37
        3.3.2 Informational Metrics 37
        3.3.3 MLE Versus Other Estimators 38

		3.4 Constrained Maximum Likelihood Estimation	39
		3.4.1 Constrained Cramér-Rao Lower Bound	41
	3.5	Bayesian Estimation	42
	3.6	Conclusions	44
	References		44
4	**Vectorial Optics**		**47**
	4.1	Electromagnetism and Optics	47
	4.2	Polarisation of Light	49
		4.2.1 Lissajous Diagrams and the Polarisation Ellipse	51
		4.2.2 Jones Vectors	51
		4.2.3 Coherency Matrices	53
		4.2.4 Stokes Vectors	55
		4.2.5 The Poincaré Sphere	57
		4.2.6 Stokes Space	57
	4.3	Jones and Mueller Polarisation Algebras	58
	4.4	Vectorial Ray-Tracing	62
	4.5	Focusing of Vectorial Beams	62
		4.5.1 Scaled Debye-Wolf Diffraction Integral	63
		4.5.2 Focusing of Partially Polarised, Partially Coherent Light	69
		4.5.3 Examples	76
	4.6	Conclusions	81
	References		82
5	**Information in Polarimetry**		**87**
	5.1	Polarimetry	87
		5.1.1 Stokes Polarimetry	88
		5.1.2 Mueller Polarimetry	89
	5.2	Polarisation Resolution	90
		5.2.1 Polarisation Encoding and Degrees of Freedom	92
		5.2.2 Efficiency of Observation	93
		5.2.3 Examples	95
		5.2.4 Channel Capacity and Detector Numbers	100
	5.3	Optimisation of Polarimeters	101
		5.3.1 Examples	103
		5.3.2 Extension of Optimisation Results	108
	5.4	Noise Propagation in Lu–Chipman Decomposition	109
		5.4.1 Single Element Systems	109
		5.4.2 Composite Systems	110
	5.5	Conclusions	112
	References		113

Contents

6 Information in Polarisation Imaging ... 117
- 6.1 Introduction ... 117
- 6.2 Eigenfunction Expansion of the Debye-Wolf Diffraction Integral ... 118
  - 6.2.1 Derivation of the Eigenfunction Expansion ... 119
  - 6.2.2 Properties of the Eigenfunction Expansion ... 122
  - 6.2.3 Numerical Examples ... 124
- 6.3 Inversion of the Debye-Wolf Diffraction Integral ... 127
  - 6.3.1 Some Notes on Inversion ... 129
  - 6.3.2 Examples ... 132
- 6.4 Polarisation Microscopy ... 138
  - 6.4.1 Fisher Information in Microscopy ... 138
  - 6.4.2 Examples ... 141
- 6.5 Physical Constraints in Vectorial Imaging ... 154
- 6.6 Conclusions ... 157
- References ... 159

7 Multiplexed Optical Data Storage (MODS) ... 163
- 7.1 Electromagnetic Scattering Calculations ... 164
  - 7.1.1 System Description ... 165
  - 7.1.2 Mode Expansion Theory ... 166
- 7.2 Optical Disc Readout: Numerical Results ... 172
  - 7.2.1 Field Distributions Within the Data Pit ... 172
  - 7.2.2 Properties of the Scattered Field ... 174
- 7.3 Optimal Pit Geometry ... 178
- 7.4 Conclusions ... 180
- References ... 181

8 Single Molecule Studies ... 183
- 8.1 Introduction ... 183
- 8.2 Photon Statistics in Single Molecule Orientational Imaging ... 184
  - 8.2.1 Signal-to-Noise Considerations ... 184
  - 8.2.2 Probability Density Function of the Number of Detected Photons ... 186
  - 8.2.3 Probability Density Function of Time Averaged Intensity ... 187
  - 8.2.4 Three Dimensional Dipole Wobble ... 192
  - 8.2.5 Discussion ... 193
- 8.3 Longitudinal Dipole Orientation and Field Mapping ... 194
  - 8.3.1 Description of System ... 194
  - 8.3.2 System Tolerances ... 198
- 8.4 Conclusions ... 202
- References ... 202

9 Conclusions	205
Appendix A: Some Information Theoretic Proofs	211
Appendix B: Special Functions	215

Abbreviations and Acronyms

nD	n-dimensional
BCRLB	Bayesian Cramér-Rao lower bound
BD	Blu-ray disc
BFIM	Bayesian Fisher information matrix
BIM	Boundary integral method
CCD	Charge-coupled device
CCRLB	Constrained Cramér-Rao lower bound
CD	Compact disc
CDF	Cumulative distribution function
CMLE	Constrained maximum likelihood estimator
CRLB	Cramér-Rao lower bound
CSDM	Cross-spectral density matrix
DOAP	Division of amplitude polarimeter
DOWP	Division of wavefront polarimeter
DVD	Digital versatile disc
EDF	Extended depth of field
FDTD	Finite difference time domain
FEM	Finite element method
FIM	Fisher information matrix
GFP	Green fluorescent protein
HD-DVD	High density digital versatile disc
MAP	Maximum a *posteriori*
ML	Maximum likelihood
MLE	Maximum likelihood estimator
MODS	Multiplexed optical data storage
NA	Numerical aperture
OCT	Optical coherence tomography
ODS	Optical data storage
PDF	Probability density/distribution function
PSA	Polarisation state analyser
PSG	Polarisation state generator

RAM	Random access memory
RMS	Root mean square
SIL	Solid immersion lens
SLM	Spatial light modulator
SMD	Single molecule detection
SNR	Signal to noise ratio
STED	Stimulated emission depletion microscopy
TE	Transverse electric
TM	Transverse magnetic
WDM	Wavelength division multiplexing
WSS	Wide-sense stationary

Chapter 1
Introduction

> *Science is facts; just as houses are made of stones, so is science made of facts; but a pile of stones is not a house and a collection of facts is not necessarily science.*
>
> J. Henri Poincaré

1.1 Motivation and Aims

Light, or rather optics, has provided the means to learn and gather information about the physical world throughout history. The inexorable march of science and technology, has for example, seen the development of the telescope, microscope, camera, optical fibre and laser, to mention but a few. In a world where science moves to ever smaller scales and more specialised problems however, the boundaries of current technology are continually challenged, motivating the search for more sophisticated systems providing greater information content, sensitivity and increased dimensionality.

Traditionally, such advances arise from new or refined theories and techniques, as exemplified by the success of quantum optics. Quantum optics has, for example, facilitated resolution gains in imaging by means of squeezed light, or more exotically, allowed so-called "ghost imaging" in which entangled photons are used to image an object indirectly (see [8] for a short review).

Alternatively, utilising previously neglected, or by introducing additional, degrees of freedom can allow further progress. Spectroscopic studies, for instance, inherently possess additional information carrying channels in the form of multiple wavelengths and are hence frequently conducted. Full electromagnetic treatments of optical systems are also often eschewed in favour of simpler scalar methods yet the vectorial nature of light need not be considered a hindrance, but instead can be viewed as affording the additional dimensionality sought. Despite the associated modelling

M. R. Foreman, *Informational Limits in Optical Polarimetry and Vectorial Imaging*, Springer Theses, DOI: 10.1007/978-3-642-28528-8_1,
© Springer-Verlag Berlin Heidelberg 2012

and mathematical complexities, polarisation represents a promising candidate for the solution of a number of current problems. This thesis is therefore dedicated to exploring the capabilities and opportunities afforded by polarised light.

Development of polarisation based optical systems can pose significant difficulties in modelling. Vectorial propagation, using for instance the Stratton-Chu integral [11] or a Green's tensor formulation [12], are frequently cumbersome in both a mathematical and computational respect. One objective of the research detailed in this thesis thus strives to mitigate such modelling burdens, whilst also attempting to gain an understanding of the behaviour and properties of polarised light in optical systems. The tools and physical insight gained will prove applicable to optical data storage, lithography, microscopy and many other fields [4, 6, 10].

Whilst the polarisation state of light itself, can be used to transmit information, hence presenting new possibilities in communications and optical data storage, changes in polarisation induced by an object can, alternatively, be used to expose new material or sample properties to scientific scrutiny, such as the orientation of fluorescent molecules [2]. Studies of this nature however necessitate detection architectures from which the polarisation state of light can be found. Polarimetry is a mature field [1], nevertheless, as with all empirical data, measurements are inevitably corrupted by the presence of noise and other unknown perturbations. Fundamentally the resulting uncertainty sets a limit on the achievable accuracy and hence the obtainable information. Quantification of these performance limits is important not only for the comparison of alternative detection schemes, but also for their refinement and optimisation. So doing hence allows the potential of polarisation based systems to be fully exploited, and therefore constitutes a further objective of this research.

Conventional metrics, such as signal to noise ratios and resolution, are unfortunately less appropriate in a polarisation domain, since they refer to irradiance measurements. Informatic and statistical principles, e.g. maximum likelihood estimators [9], which adopt a more task specific perspective, will hence be applied. The generality afforded by an information based approach will be seen to permit application to a large array of different polarisation based systems, intended for quite disparate purposes, but also allow the incorporation of any a priori knowledge with regards to the system into optimisation routines [3].

General formulations, although invaluable, do not immediately reveal potential technological advances. A final goal of this thesis is hence to examine current problems which can be addressed using polarised light. Polarisation microscopy, for example, represents the natural fusion of existing imaging techniques with polarisation based measurements and is one example that will be considered. Furthermore the need to store and transport vast volumes of data is growing. Optical data storage (ODS) has become a well established solution, however conventional means by which to further increase storage capacity are reaching their limits. A polarisation based multiplexed optical data storage solution in which multiple bits can be encoded into a single data pit will hence be examined and optimised.

As a final example, attention is given to single molecule studies, in which polarisation can play an important role, e.g. for orientational measurements. Such studies can for example have distinct noise properties, which must be considered to achieve

1.1 Motivation and Aims

high accuracy measurements [5]. Furthermore, whilst current techniques are limited to measurement of the transverse orientation, a new system is presented in which longitudinal measurements can be made, hence opening the way to 3D polarimetry and the additional information such studies can provide.

Ultimately it is hoped that the work presented in this thesis provides an insight into the capabilities and advances afforded by a number of specific polarisation based optical systems. Although exhaustive examination of all such systems lies far beyond the scope of this work, the tools and techniques also developed herein are intended to be applicable to the modelling and analysis of alternative problems.

1.2 Thesis Structure and Overview

Chapters 2 and 3 provide an introduction to key concepts taken from probability, information and estimation theory which will be routinely used throughout this thesis. Such principles are important in describing and parameterising noise present in an optical system which fundamentally limits the information obtainable in measurement and parameter inference. Key estimation routines, such as maximum likelihood and Bayesian estimation will hence be discussed, allowing precision limits to be defined and studied quantitatively.

Chapter 4 continues by providing a theoretical introduction to vectorial optics, starting first with a short discussion of Maxwell's equations. Numerous alternative descriptions of polarised light are subsequently presented, again due to their common use during the course of this work. Such discussions will naturally lead to Jones and Mueller calculus: tools which are invaluable for vectorial ray tracing (an extension of the more familiar scalar ray tracing methods), in turn aiding analysis of complex optical systems. Finally, due to the ubiquity of lenses in modern day optics the chapter concludes with a discussion of the scaled Debye-Wolf integral, which can be used to describe the focusing of polarised light under a wide range of circumstances. In particular a new formalism by which spatially inhomogeneous partially coherent, partially polarised light can be focused is detailed and illustrated by a number of important numerical examples.

Chapter 5 considers the theories of Chaps. 2–4 and applies them to the field of polarimetry. Polarimetry aims at measuring the state of polarisation of light and/or the polarisation changing properties of a sample. This chapter considers both types of polarimetry and gives a framework within which the measurement systems can be optimised, accounting for signal dependent noise and incorporation of any a priori information that may be held. The optimisation procedure is highlighted by means of a number of examples, before discussion of how the algorithm can be extended to more complex inference problems is given. In particular the chapter concludes with a discussion of noise propagation in the ever-popular Lu-Chipman decomposition [7].

Chapter 6 extends the ideas of Chap. 5 by considering the growing field of polarisation microscopy. In an attempt to improve the imaging performance of polarisation microscopes an eigenfunction analysis of focusing by a high numerical aperture lens is derived and applied to a number of inverse problems of significance in microscopy, such as superresolution. Issues pertaining to this inversion procedure are also fully discussed. Consideration is further given to system performance and potential crosstalk when imaging single or multiple dipole sources. Finally, since modal analysis of optical systems does not necessarily ensure physicality, in the sense that basis modes do not satisfy Maxwell's equations, the question as to how such physical constraints can be built into the estimation procedure, and the potential accuracy improvements, is evaluated.

Having developed a number of general informational theories and tools in the earlier part of this thesis, Chaps. 7 and 8 consider two specific systems. Chapter 7 considers a novel optical data storage system in which polarisation encoding is used to store large volumes of data on a single optical disc. In particular the system will be fully modelled including the scattering of polarised light from the disc surface. Numerical simulations allow the data pit dimensions to be optimised and results in this vein are given.

Chapter 8 on the other hand considers polarisation based single molecule experiments. In this context a specific noise model is developed accounting for random orientational changes of molecules which may arise. Finally a novel system capable of measuring the full three-dimensional orientation of single molecules in real time is introduced. This is in contrast to existing systems, which measure the transverse orientation only. Key tolerances of the system are also discussed.

Finally in Chap. 9 the main results and ideas of this thesis are collated and discussed. Opportunities for further work and areas of development are also highlighted.

References

1. R.M.A. Azzam, N.M. Bashara, *Ellipsometry and Polarised Light* (Elsevier, Amsterdam, 1987)
2. M.R. Foreman, C. Macías Romero, P. Török, Determination of the three dimensional orientation of single molecules. Opt. Lett. **33**, 1020–1022 (2008)
3. M.R. Foreman, C. Macías Romero, P. Török, A priori information and optimisation in polarimetry. Opt. Express **16**, 15212–15227 (2008)
4. M.R. Foreman, S.S. Sherif, P.R.T. Munro, P. Török, Inversion of the Debye-Wolf diffraction integral using an eigenfunction representation of the electric fields in the focal region. Opt. Express **16**, 4901–4917 (2008)
5. M.R. Foreman, S.S. Sherif, P. Török, Photon statistics in single molecule orientational imaging. Opt. Express **15**, 13597–13606 (2007)
6. M.R. Foreman, P. Török, Focusing of spatially inhomogeneous partially coherent, partially polarized electromagnetic fields. J. Opt. Soc. Am. A **26**, 2470–2479 (2009)
7. S.Y. Lu, R.A. Chipman, Interpretation of Mueller matrices based on polar decomposition. J. Opt. Soc. Am. A **13**, 1106–1113 (1996)
8. L.A. Lugiato, A. Gatti, E. Brambilla, Quantum imaging. J. Opt. B: Quantum Semiclass. Opt. **4**, 176–183 (2002)
9. L.L. Scharf, *Statistical Signal Processing: Detection, Estimation, and Time Series Analysis* (Addison-Wesley Publishing Co., Reading, 1991)
10. S.S. Sherif, M.R. Foreman, P. Török, Eigenfunction expansion of the electric fields in the focal region of a high numerical aperture focusing system. Opt. Express **16**, 3397–3407 (2008)

References

11. J.A. Stratton, L.J. Chu, Diffraction theory for electromagnetic waves. Phys. Rev. **56**, 99–107 (1939)
12. A. S. van de Nes, Rigorous electromagnetic field calculations for advanced optical systems. Ph.D. thesis, Delft University of Technology (2005)

Chapter 2
Fundamentals of Probability Theory

> *The scientist has a lot of experience with ignorance and doubt and uncertainty, and this experience is of very great importance, I think.*
>
> Richard P. Feynman

Probability is a concept familiar to the vast majority of readers on an intuitive level, however in a stricter sense it is generally poorly understood. A substantial fraction of this thesis draws ideas and tools from the rigorous theories of probability and thus it is apt to provide a fuller account of the rudimentary mathematical principles. This chapter does not aim to give an exhaustive exposition of probability theory (fuller discussion can be found in many existing texts, such as [2, 5, 10, 11, 16]), but instead particular attention is given to the meaning of randomness, whereby a suitable parameterisation of the statistics involved can be formulated.

Ultimately, it is the uncertainty so described which, in its various forms, simultaneously allows "information" to be gained from observations of a system, yet also limits the confidence held in those observations [24–26, 29]. An understanding of the ideas presented here will therefore prove invaluable later, not only for describing many electromagnetic sources which are inherently stochastic (see Chaps. 4 and 8), but also when considering bounds in optical systems (Chaps. 3–6), and how to overcome these limits (Chaps. 5–7). Finally, this chapter serves as a platform from which to establish the mathematical notation which will be used throughout this text.

2.1 Random Variables

2.1.1 Specifying a Random Variable

2.1.1.1 Cumulative and Probability Distribution Functions

Central to probability theory are the notions of random experiments and random variables. The former, defined as an experiment in which the result varies even when performed under identical conditions, can not be described using deterministic principles. Instead a probability is ascribed to each possible outcome, describing the relative frequency with which they occur.[1] A random variable X can then be rigorously defined as a quantity for which a real number x is assigned dependent on the outcome of a random experiment according to a fixed, deterministic rule. Based upon an experiment in which the number of photons are counted on two different detectors, for example, it is possible to define an assortment of different random variables such as the sum or difference of the photon counts measured. It should be noted that, as has been indicated here, uppercase letters will be used to denote a random variable, whilst the lowercase counterpart indicates a particular value taken by that variable.

Specification of random variables, both fully and approximately, serves as an important first step in many statistical problems. Discrete random variables can be fully specified by the denumeration of all possible values in conjunction with the associated probabilities, $p_X(x) = p(X = x)$, or relative frequency ($0 \leq p_X(x) \leq 1$). Continuous variables, for example the length of time between arrivals of photons at a detector, however can not be described in this way. Instead the probability, $F_X(x)$, that the random variable lies below a given value, i.e. $X \leq x$, can be given for all possible values of x. $F_X(x)$ is known as the cumulative distribution function (CDF) of the random variable X. It should be apparent that a discrete random variable can also be fully described using a CDF.

Using the CDF it is further possible to calculate the probability that the value of a random variable lies within a given range, say between x and $x + \delta x$, according to

$$p(x < X \leq x + \delta x) = F_X(x + \delta x) - F_X(x), \tag{2.1}$$

or alternatively the average "density" of probability within this range

$$\frac{p(x < X \leq x + \delta x)}{\delta x} = \frac{F_X(x + \delta x) - F_X(x)}{\delta x}. \tag{2.2}$$

Taking the limit of an infinitesimal range, i.e. $\delta x \to 0$, yields an alternative representation of a random variable, known as a probability density/distribution function (PDF) viz.

$$f_X(x) = \lim_{\delta x \to 0} \left[\frac{F_X(x + \delta x) - F_X(x)}{\delta x} \right] = \frac{dF_X(x)}{dx}. \tag{2.3}$$

Probability density functions provide a much more useful specification of a random variable and it will be seen that further reference to CDFs will be limited. By way of comparison Fig. 2.1 shows the CDF and PDF for two types of random variables

---

[1] Although relative frequency, defined as $f = \lim_{N \to \infty} n/N$, where n is the number of occurrences of a particular outcome in N repetitions of the random experiment, is perhaps the most intuitive manner in which to define probabilities, an axiomatic definition as first formulated by Kolmogorov [14, 15], is sometimes preferred since there is no guarantee that this limit exists.

2.1 Random Variables

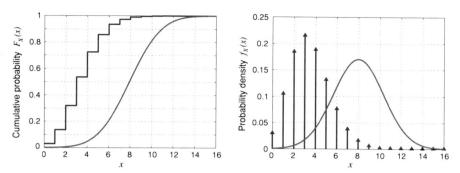

Fig. 2.1 CDFs (*left*) and PDFs (*right*) for a Poisson random variable (*blue*) with a mean and variance of 3.5, and a Gaussian random variable (*green*) with a mean of 8 and variance of 5.5

that will be important in this work, namely a (continuous) Gaussian and (discrete) Poisson random variable (see Sect. 2.3 for a fuller introduction to these variables). With reference to Fig. 2.1 it should be noted that defining a PDF for discrete random variables requires the introduction of the Dirac delta function, denoted $\delta(x - x_0)$, whereby

$$f_X(x) = \sum_j p_X(x_j)\delta(x - x_j), \tag{2.4}$$

where x_j denotes the possible discrete values of X.

2.1.1.2 Characteristic Functions and Moments of a Random Variable

Frequently Fourier analysis can afford both insight and computational advantages when considering complex systems. Such a statement can also be made when considering statistical problems, especially when considering sums of independent random variables in which the requisite convolution of PDFs reduces to a simple product [5]. Motivated by these potential advantages consider the Fourier transform of a PDF $f_X(x)$ defined by

$$\mathcal{F}[f_X(x)] = \Phi_X(\omega) = \int_{-\infty}^{\infty} f_X(x)\exp(i\omega x)dx, \tag{2.5}$$

$$= E\left[\exp(i\omega X)\right], \tag{2.6}$$

commonly known as the characteristic function of X. Equation (2.6) gives an alternative interpretation of the characteristic function as the expected value, denoted $E[\cdots]$, of a function of X, namely $\exp(i\omega X)$, in which ω is viewed as a free parameter. A power series expansion of the exponential term yields

$$\Phi_X(\omega) = \int_{-\infty}^{\infty} f_X(x) \left[1 + i\omega x + \frac{(i\omega x)^2}{2!} + \cdots + \frac{(i\omega x)^n}{n!} + \cdots \right] dx, \quad (2.7)$$

$$= 1 + i\omega E[X] + \frac{(i\omega)^2}{2!} E[X^2] + \cdots + \frac{(i\omega)^n}{n!} E[X^n] + \cdots, \quad (2.8)$$

where it has been assumed that the order of summation and integration can be interchanged (requiring all terms to be finite and convergence of the resulting series [7]). The expectation

$$E[X^n] = \int_{-\infty}^{\infty} x^n f_X(x) dx \quad (2.9)$$

is known as the nth order moment of X and from Eq. (2.7) it can be seen that given knowledge of the infinite set of moments of a random variable it is possible to calculate the PDF and hence completely describe the random variable.

Practically speaking it is unfeasible to calculate the full set of nth order moments, however knowledge of the first few orders can prove useful in parameterising the salient properties. Of particular interest are the first and second order moments, $E[X]$ and $E[X^2]$. The former, also known as the arithmetic mean, describes the central value about which the random variable X varies, whilst the latter can be used to calculate the spread of variation, or variance, about the mean value via

$$\text{VAR}[X] = E[(X - E[X])^2] = E[X^2] - E[X]^2. \quad (2.10)$$

The variance is sometimes also called the second order centered moment of the random variable X. Higher order centered moments can be defined in an analogous way, providing an alternative parameterisation.

The mean and variance of a random variable are particularly important parameters partly because Gaussian random variables are widespread in physics, as follows from the Central Limit Theorem (see Sect. 2.3). Owing to their symmetry, Gaussian random variables are fully specified with knowledge of only the first two moments and hence are often used as a good first order approximation in many problems.

In closing, it is important to mention that although the above analysis has been performed in terms of a Fourier transform of the PDF, for discrete random variables a z-transform as defined by

$$\mathcal{Z}[p_X(x)] = G_X(z) = \sum_{x=0}^{\infty} p_X(x) z^x = E[z^X] \quad (2.11)$$

is more appropriate, whilst for positive valued random variables the Laplace transform defined by

2.1 Random Variables

$$\mathcal{L}[f_X(x)] = X^*(s) = \int_0^\infty f_X(x)\exp(-sx)dx = E[\exp(-sx)] \qquad (2.12)$$

should be used [16].

2.1.2 Functions of a Random Variable

Mathematical transformations are regularly applied to random variables. A suitable characterisation of the resulting variable, which is itself also random then often becomes necessary. If the random variable Y is defined via the transformation $Y = g(X)$ it is possible to calculate the CDF and PDF of Y from those specifying X. To illustrate how this can be done, consider an arbitrary nonlinear function $y = g(x)$ as shown in Fig. 2.2. From this figure it can be seen that if Y lies in the range $y < Y \leq y + \delta y$ it is equivalent to the untransformed variable X having a value in the range $x_1 < X \leq x_1 + \delta x_1$ or $x_2 < X \leq x_2 + \delta x_2$. The probabilities of these two equivalent events occurring can be found approximately using Eqs. (2.2) and (2.3) and are given by

$$p(y < Y \leq y + \delta y) \approx f_Y(y)\delta y \qquad (2.13)$$

and

$$p(\{x_1 < X \leq x_1 + \delta x_1\} \cup \{x_2 < X \leq x_2 + \delta x_2\}) \approx f_X(x_1)\delta x_1 + f_X(x_2)\delta x_2 \qquad (2.14)$$

respectively, where $\cup$ denotes the union of two sets. Since the two events are equivalent the probabilities must be equal. Equating Eqs. (2.14) and (2.13) and performing the limiting process, $\delta y \to 0$, such that the approximate expressions become exact, therefore yields the desired result

$$f_Y(y) = \sum_k \left.\frac{f_X(x)}{|dy/dx|}\right|_{x=x_k}, \qquad (2.15)$$

where x_k are the solutions to the equation $y = g(x)$.

2.1.3 Multiple Random Variables

Analysis of random behaviour in a practical system frequently requires dealing with multiple random variables. This is for example vital in image processing, wherein the intensity recorded on each pixel of a charge-coupled device (CCD)

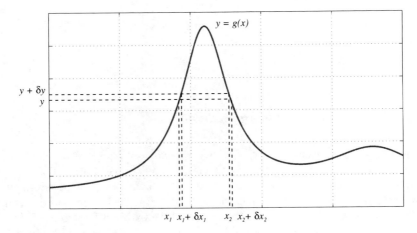

Fig. 2.2 The equivalent event for $y \leq Y \leq y + \delta y$ under the arbitrary transformation $y = g(x)$ is shown to be $x_1 < X \leq x_1 + \delta x_1$ or $x_2 < X \leq x_2 + \delta x_2$

is subject to different noise during the image capture process and hence requires a multivariate approach, see for example [4, 27]. Depending on the noise source, noise in neighbouring pixels may exhibit some degree of dependence on each other and it is furthermore important to be able to quantify such properties. It is these issues which are addressed in this section.

2.1.3.1 Joint, Conditional and Marginal Distribution Functions

Letting $X_1, X_2, \ldots, X_n$ denote a collection of n random variables it is a simple matter to extend the results of Sects. 2.1.1 and 2.1.2 to describe their joint behaviour. In doing so it will be convenient to collect all random variables together to form a vector random variable denoted $\mathbf{X} = (X_1, X_2, \ldots, X_n)$. The joint CDF, $F_{\mathbf{X}}(\mathbf{x}) = F_{\mathbf{X}}(x_1, x_2, \ldots, x_n)$, can then be defined as the probability that $\mathbf{X} \leq \mathbf{x}$ where the inequality holds elementwise, i.e. $\{X_1 \leq x_1, X_2 \leq x_2, \ldots, X_n \leq x_n\}$. Furthermore the joint PDF can be defined analogously to Eq. (2.3) as

$$f_{\mathbf{X}}(\mathbf{x}) = \frac{\partial^n F_{\mathbf{X}}(\mathbf{x})}{\partial x_1 \partial x_2 \ldots \partial x_n}. \qquad (2.16)$$

The random behaviour of a single random variable, say X_j, in isolation may however still be of interest. In essence, this means considering the probability of X_j adopting a particular value x_j irrespective of the values of all other random variables X_k, where $k \neq j$. Since by construction x_k must adopt a value in the range $-\infty < x_k \leq \infty$ the marginal CDF of X_j, which fully specifies the statistics of X_j alone, is given by $F_{X_j}(x_j) = F_{\mathbf{X}}(\infty, \ldots, x_j, \ldots, \infty)$. By differentiation, a set of marginal PDFs can also be found, which are given by

2.1 Random Variables

$$f_{X_j}(x_j) = \int_{-\infty}^{\infty} \cdots \int_{-\infty}^{\infty} f_{\mathbf{X}}(x_1', \ldots, x_j, \ldots, x_n') dx_1' \ldots dx_{j-1}' dx_{j+1}' \ldots dx_n'. \tag{2.17}$$

A final distribution function which proves useful in multivariate problems is the conditional CDF and/or PDF. A conditional probability describes the relative frequency of a particular event occurring given that some condition is known to be satisfied. For example, in a noisy binary communication channel either a logical 0 or 1 are transmitted, however the receiver is subject to random noise, meaning the raw detected signal can assume a continuous range of values. In this case the probability that the transmitted message was a 0 (or a 1) given the raw recorded data is of interest [1].

The conditional probability, $p(B|A)$, of an event B occurring given that an event A has occurred can be shown, via a relative frequency argument [7], to obey the relation

$$p(A, B) = p(B|A) p(A), \tag{2.18}$$

where $p(A, B)$ and $p(B|A)$ are the joint and conditional probabilities respectively. Using Eq. (2.18) conditional PDFs can be derived, for example the conditional PDF of X_n given the values of $X_1, X_2, \ldots X_{n-1}$, is given by

$$f_{X_n}(x_n | x_1, x_2, \ldots, x_{n-1}) = \frac{f_{X_1, X_2, \ldots, X_n}(x_1, x_2, \ldots, x_n)}{f_{X_1, X_2, \ldots, X_{n-1}}(x_1, x_2, \ldots, x_{n-1})}. \tag{2.19}$$

Equation (2.19) can be iterated to give an alternative expression for the joint PDF

$$f_{\mathbf{X}}(\mathbf{x}) = f_{X_n}(x_n | x_1, \ldots, x_{n-1}) f_{X_{n-1}}(x_{n-1} | x_1, \ldots, x_{n-2}) \ldots f_{X_2}(x_2 | x_1) f_{X_1}(x_1). \tag{2.20}$$

Finally it remains to consider functions of multiple random variables. The analog to Eq. (2.15) is all that is required, which is given by

$$f_{\mathbf{Y}}(\mathbf{y}) = \frac{f_{\mathbf{X}}(h_1(\mathbf{y}), h_2(\mathbf{y}), \ldots, h_n(\mathbf{y}))}{|\mathbb{J}(\mathbf{x})|}, \tag{2.21}$$

where $Y_1 = g_1(\mathbf{X}), Y_2 = g_2(\mathbf{X}), \ldots, Y_n = g_n(\mathbf{X})$ and

$$x_1 = h_1(\mathbf{y}), \quad x_2 = h_2(\mathbf{y}), \quad \ldots \quad x_n = h_n(\mathbf{y}), \tag{2.22}$$

are the unique solutions to the set of equations

$$y_1 = g_1(\mathbf{x}), \quad y_2 = g_2(\mathbf{x}), \quad \ldots \quad y_n = g_n(\mathbf{x}), \tag{2.23}$$

and $|\mathbb{J}(\mathbf{x})|$ is the determinant of the Jacobian,

$$\mathbb{J}(\mathbf{x}) = \begin{pmatrix} \frac{\partial g_1}{\partial x_1} & \cdots & \frac{\partial g_1}{\partial x_n} \\ \vdots & \ddots & \vdots \\ \frac{\partial g_n}{\partial x_1} & \cdots & \frac{\partial g_n}{\partial x_n} \end{pmatrix}, \qquad (2.24)$$

of the transformation. This result can be derived using a multi-dimensional extension to the proof of Sect. 2.1.2 [16].

2.1.3.2 Cross-Correlation Matrices and Statistical Independence

In Sect. 2.1.1.2 the concept of (centered) moments of a random variable was introduced, with particular emphasis given to the first and second order moments. Whilst for a single random variable it was found to be important to parameterise the extent of variation, it can frequently be equally important to describe how closely the variations in one random variable X_1 follow those of another X_2 as can be described by the second order joint moment (correlation), defined as

$$\text{COR}[X_1, X_2] = E[X_1 X_2], \qquad (2.25)$$

or the second order centered joint moment (covariance)

$$\text{COV}[X_1, X_2] = E[(X_1 - E[X_1])(X_2 - E[X_2])]. \qquad (2.26)$$

For zero mean random variables the correlation and covariance are equivalent and so these terms will often be used interchangeably throughout this work when appropriate. Variance and covariance properties of multiple random variables can conveniently be combined into a cross-correlation or covariance matrix defined respectively as

$$\mathbb{C} = E[\mathbf{X}\mathbf{X}^T], \qquad (2.27)$$
$$\mathbb{K} = E[(\mathbf{X} - E[\mathbf{X}])(\mathbf{X} - E[\mathbf{X}])^T], \qquad (2.28)$$

where T denotes matrix transposition. The jth diagonal term of the correlation (or covariance) matrix gives the second order moment (variance) of X_j, whilst the $\{j, k\}$th off diagonal element describes the correlation (covariance) between X_j and X_k.

Covariance of two variables describes the extent to which there is a linear dependence between their values, however a zero covariance does not imply that there is no relationship. A stricter requirement on two random variables is that of statistical independence, which demands that the conditional PDF of X_j given X_k be functionally invariant to X_k i.e. $f_{X_j}(x_j|x_k) = f_{X_j}(x_j)$. As a result it is possible to give conditions for two or more random variables $X_1, X_2, \ldots, X_n$ to be independent [16], namely

2.1 Random Variables

$$F_{\mathbf{X}}(\mathbf{x}) = \prod_{k=1}^{n} F_{X_k}(x_k) \quad \text{for all } \mathbf{x}. \tag{2.29}$$

For discrete and continuous random variables this can be expressed in the form $p_{\mathbf{X}}(\mathbf{x}) = \prod_{k=1}^{n} p_{X_k}(x_k)$ and $f_{\mathbf{X}}(\mathbf{x}) = \prod_{k=1}^{n} f_{X_k}(x_k)$ respectively, for all $\mathbf{x}$. Such criteria can prove useful in identifying when variables can be considered separately and in gaining insight into inherent performance trade-offs in physical systems which may exist. An example of this will be presented in Chap. 5.

2.1.3.3 Complex Random Variables

Hitherto discussion has been limited to real valued random variables, however in electromagnetism and optics many stochastic quantities, such as the electric field vector, are complex valued and as such it is necessary to consider how to mathematically describe these cases. Fortunately the essential tools require no further development since, if it is noted that a single complex valued random variable has two degrees of freedom, results pertaining to multiple random variables can be applied. There are a number of alternatives approaches in the literature [7, 28] as to which degrees of freedom to treat as random variables. One option is to treat the real and imaginary parts as two random variables. Specification of a complex, random variable $\mathbf{X} = \mathbf{U} + i\mathbf{V}$ can then be achieved, for example, by provision of the joint PDF of its real and imaginary parts $f_{\mathbf{X}}(\mathbf{x}) = f_{\mathbf{U},\mathbf{V}}(\mathbf{u}, \mathbf{v})$. Depending on the problem at hand, it may however be more convenient to use the modulus and argument (amplitude and phase) of $\mathbf{X}$ or alternatively $\mathbf{X}$ and its complex conjugate $\mathbf{X}^*$.

Characterisation of complex random variables is again afforded by correlation and covariance matrices, however these can now be defined in two ways. Firstly, if considering the real and imaginary part of the complex vector random variable as two distinct vector random variables, and defining a new real vector random variable $(\mathbf{U}, \mathbf{V})$ the correlation and covariance matrices are defined as per Eqs. (2.27) and (2.28) (similarly if considering the amplitude and phase). Alternatively complex matrices can be defined (under an assumption of circularity [7]) according to

$$\mathbb{C} = E[\mathbf{X}\mathbf{X}^\dagger], \tag{2.30}$$

$$\mathbb{K} = E[(\mathbf{X} - E[\mathbf{X}])(\mathbf{X} - E[\mathbf{X}])^\dagger], \tag{2.31}$$

where † denotes the Hermitian adjoint operation. These complex forms of the correlation and covariance matrices are smaller in dimension than the former definition, yet via considering their real and imaginary parts contain the same parameterisation of a complex random variable and are hence often preferred, see e.g. [12].

2.2 Random Processes

Random variables need not be limited to a single scalar (or vector) quantity, but could feasibly be a function of both space and time, whereby they are designated as a random, or stochastic process. An illustrative example is that of imaging laser light scattered from a rough surface. Due to the spatial variations in optical path length seen by different scattered parts of the laser beam the scattered distribution exhibits strong spatial intensity fluctuations, known as a speckle pattern [7]. Furthermore, if the scattered light is passed through a time varying element, such as a turbulent medium the speckle pattern is seen to vary in time [8], due to the changes in effective refractive index and hence optical path length through the medium. Although, given an exact surface profile and knowledge of the spatial and velocity distributions of particles comprising the turbulent media, it would be possible to deterministically calculate the scattered field/intensity distribution as a function of space and time, a statistical approach is often used to achieve a phenomenological description. Such an approach is justified since it is unlikely that any two surfaces will have the same topology or that two media will have the same particulate distributions. That said it should not be thought that statistical approaches are only used in optical fields to overcome ignorance of the exact configuration, since many processes, such as the spontaneous emission of photons by excited atoms, are inherently random. Accordingly this section considers the specification and parameterisation of random processes.

2.2.1 Specifying a Random Process

A random process $U(\mathbf{r}, t)$ can be completely specified by cataloguing all the possible sample functions $u(\mathbf{r}, t)$ (also known as realisations) and the associated probabilities with which they occur. Here $\mathbf{r}$ and t denote spatial and temporal coordinates respectively. Once again it is highly impractical to specify a random process by enumerating all possible realisations, and may in fact be impossible if the realisations can not be written in a functional form. An alternative specification is offered by the full set of kth order joint PDFs $f_\mathbf{U}(\mathbf{u}_1, \ldots, \mathbf{u}_k; \mathbf{r}_1, \ldots, \mathbf{r}_k; t_1, \ldots, t_k)$, where $\mathbf{u}_j$ is a sample of the complex vector random process $\mathbf{U}(\mathbf{r}, t)$ at position $\mathbf{r}_j$ and time t_j. Characterisation of the significant statistical behaviour of a random process can however often be adequately described using only the first and second order joint PDFs and henceforth consideration will only be given to these.

2.2.2 Stationarity and Ergodicity

Much of deterministic physics examines steady state solutions of systems. Similarly in statistical physics the vast majority of work considers random processes whose

2.2 Random Processes

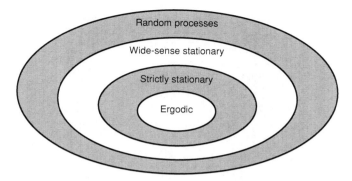

Fig. 2.3 Classes of random processes. Figure based on Fig. 3.4 of [7]

properties do not vary in time. Such random processes are termed (strictly) stationary, which formally implies that all kth order joint PDFs are invariant under arbitrary shifts in time, τ, i.e.

$$f_{\mathbf{U}}(\mathbf{u}_1, \ldots, \mathbf{u}_k; \mathbf{r}_1, \ldots, \mathbf{r}_k; t_1, \ldots, t_k)$$
$$= f_{\mathbf{U}}(\mathbf{u}_1, \ldots, \mathbf{u}_k; \mathbf{r}_1, \ldots, \mathbf{r}_k; t_1 - \tau, \ldots, t_k - \tau), \quad (2.32)$$

for all τ. As a result the first order PDF is independent of time, whilst the second order PDF, depends only on the time difference $\tau = t_1 - t_2$ between sample points. Consequently the mean,

$$E[\mathbf{U}(\mathbf{r}, t)] = \boldsymbol{\mu}(\mathbf{r}), \quad (2.33)$$

is independent of time, whilst the correlation and covariance matrices given by the moments

$$\text{COR}[\mathbf{U}(\mathbf{r}_1, t_1), \mathbf{U}(\mathbf{r}_2, t_2)] = E[\mathbf{U}(\mathbf{r}_1, t_1)\mathbf{U}^\dagger(\mathbf{r}_2, t_2)] = \mathbb{C}(\mathbf{r}_1, \mathbf{r}_2, \tau), \quad (2.34)$$
$$\text{COV}[\mathbf{U}(\mathbf{r}_1, t_1), \mathbf{U}(\mathbf{r}_2, t_2)] = E[\{\mathbf{U}(\mathbf{r}_1, t_1) - \boldsymbol{\mu}(\mathbf{r}_1)\}\{\mathbf{U}(\mathbf{r}_2, t_2) - \boldsymbol{\mu}(\mathbf{r}_2)\}^\dagger],$$
$$= \mathbb{K}(\mathbf{r}_1, \mathbf{r}_2, \tau), \quad (2.35)$$

are functions of τ for all t_1 and t_2. The (on) off diagonal terms of these matrices are frequently termed (auto-) cross-correlations and covariances respectively.

Since specification of the infinite set of joint PDFs is not possible, it is equally difficult to establish the stationarity of a random process in general. That said, it is commonly possible to determine whether the conditions given by Eqs. (2.33–2.35) are satisfied. If they do hold for a random process then that process is said to be wide-sense stationary (WSS). Any strictly stationary process is automatically WSS, however wide-sense stationarity does not imply strict stationarity.

Common reference is often made in the literature to ergodic random processes and to conclude this section it is worthwhile defining this class of processes. An ergodic process can be defined as a random process for which any temporal average derived from a realisation of that process is the same as the ensemble average of the same quantity, i.e.

$$\langle g(U) \rangle = \lim_{T \to \infty} \frac{1}{T} \int_{-T/2}^{T/2} g[u(t)] dt = \int_{-\infty}^{\infty} g(u) f_U(u) du = E[U]. \quad (2.36)$$

From this definition it can be seen that the ergodic class of random processes is a subset of the class of strictly stationary random processes, as illustrated in Fig. 2.3. Whilst this definition of ergodicity is common in statistical optics literature (see e.g. [7]), ergodic theory constitutes a large body of research from which alternative, more general definitions and theorems originate [17].

2.2.3 Spectral Analysis of a Random Process

Random processes have thus far been discussed in a space-time domain, however motivated by the spectral methods commonly used in electromagnetism and other physical theories, it is constructive to consider a space-frequency description instead. This will prove necessary for example in Chap. 4 where coherent optical methods valid for quasi-monochromatic light only are extended to describe partially coherent systems. Initially consideration will be restricted to WSS processes, however some closing remarks will be made with regards to the relaxation of this assertion.

Spectral analysis of random processes can be performed by taking the Fourier transform of each realisation, however care must be taken since the Fourier transform of a general function need not exist. This is especially true when considering a realisation of a WSS random process, since by virtue of its stationarity the function does not decay to zero as $t \to \pm\infty$ and thus is not square integrable. Expressed more physically this means each realisation taken over all time possesses infinite energy. Consequently, performing a spectral analysis of a random process requires a more formal mathematical framework. This was provided by Wiener and Khintchine in the form of generalised harmonic analysis and stochastic Fourier-Stieltjes integrals [13, 30]. A more heuristic description of their solution in terms of integrated spectra can be found in [20]. Using the theories of Wiener and Khintchine, a rigorous meaning can be given to the Fourier integral

$$\tilde{\mathbf{u}}(\mathbf{r}, \omega) = \mathcal{F}[\mathbf{u}(\mathbf{r}, t)] = \int_{-\infty}^{\infty} \mathbf{u}(\mathbf{r}, t) \exp(i\omega t) dt, \quad (2.37)$$

allowing all previous results to be expressed in the frequency domain. Equation (2.37) must still however be used with caution, since convergence of the associated integral is not guaranteed, e.g. the phase in a white noise random process does not converge

2.2 Random Processes

over an infinite time interval. Convergence will be assumed henceforth. Of particular importance is the correlation function

$$\text{COR}[\tilde{\mathbf{U}}(\mathbf{r}_1, \omega_1), \tilde{\mathbf{U}}(\mathbf{r}_2, \omega_2)] = \int_{-\infty}^{\infty} \int_{-\infty}^{\infty} E[\mathbf{U}(\mathbf{r}_1, t_1) \mathbf{U}^\dagger(\mathbf{r}_2, t_2)] e^{i(\omega_1 t_1 - \omega_2 t_2)} dt_1 dt_2, \quad (2.38)$$

where the order of integration and expectation have been switched. Wide-sense stationarity of the process, via Eq. (2.34), gives

$$\text{COR}[\tilde{\mathbf{U}}(\mathbf{r}_1, \omega_1), \tilde{\mathbf{U}}(\mathbf{r}_2, \omega_2)] = \int_{-\infty}^{\infty} \int_{-\infty}^{\infty} \mathbb{C}(\mathbf{r}_1, \mathbf{r}_2, \tau) e^{i(\omega_1 - \omega_2) t_2} e^{-i\omega_1 \tau} dt_2 d\tau, \quad (2.39)$$

where $\tau = t_1 - t_2$. The integral over t_2 can be evaluated analytically and gives the Dirac delta function $\int_{-\infty}^{\infty} \exp(i\omega t) dt = \delta(\omega)$ thus yielding

$$\text{COR}[\tilde{\mathbf{U}}(\mathbf{r}_1, \omega_1), \tilde{\mathbf{U}}(\mathbf{r}_2, \omega_2)] = \mathbb{W}(\mathbf{r}_1, \mathbf{r}_2, \omega_1) \delta(\omega_1 - \omega_2), \quad (2.40)$$

where

$$\mathbb{W}(\mathbf{r}_1, \mathbf{r}_2, \omega) = \text{COR}[\tilde{\mathbf{U}}(\mathbf{r}_1, \omega), \tilde{\mathbf{U}}(\mathbf{r}_2, \omega)] = \int_{-\infty}^{\infty} \mathbb{C}(\mathbf{r}_1, \mathbf{r}_2, \tau) e^{-i\omega \tau} d\tau \quad (2.41)$$

is known as the cross-spectral density matrix (CSDM), which describes the correlation between harmonic components of the two complex vector random processes $\mathbf{U}(\mathbf{r}_1, t)$ and $\mathbf{U}(\mathbf{r}_2, t)$. If evaluated at the same spatial location ($\mathbf{r}_1 = \mathbf{r}_2$) Eq. (2.41) is called the Wiener-Khintchine theorem [20] and states that the spectral (power) density matrix $\mathbb{W}(\mathbf{r}, \mathbf{r}, \omega)$ and the auto-correlation matrix $\mathbb{C}(\mathbf{r}, \mathbf{r}, \tau)$ form a Fourier transform pair.

Finally for completeness a word should be given to the consequences of relaxing the assumption of a WSS process. In this case defining the concept of a spectrum proves problematic, however the dominant approach in the literature is to define time varying spectra by performing a time-windowed stationary analysis. This for example includes the Wigner-Ville, Weyl and generalised evolutionary spectra (see e.g. [9, 21–23]). Such a case will however not be considered further in this work.

2.3 Some Important Types of Statistics

A variety of different random variables and processes will be employed in the course of this work. So as to avoid the need to introduce each when required this task is performed here preemptively. Table 2.1 summarises the properties of all types of random variable that will be needed, however detailed discussion is also given here

Table 2.1 Summary of the properties of some types of random variables used during the course of this text

Variable type	Sample space	Probability/PDF	Mean $E[X]$	Variance $\mathrm{VAR}[X]$
Bernoulli	$x = \{0, 1\}$	$p_X(x) = \begin{cases} 1-p & x=0 \\ p & x=1 \end{cases}$	p	$p(1-p)$
Uniform	$x \in [a, b]$	$f_X(x) = \begin{cases} \frac{1}{b-a} & a \leq x \leq b \\ 0 & \text{otherwise} \end{cases}$	$\frac{a+b}{2}$	$\frac{(b-a)^2}{12}$
Exponential	$x \in [0, \infty]$	$f_X(x) = \begin{cases} 0 & x<0 \\ Re^{-Rx} & x \geq 0 \end{cases}$	$\frac{1}{R}$	$\frac{1}{R^2}$
Poisson	$x \in \mathbb{Z}_0^+$	$p_X(x) = \frac{\alpha^x}{x!} e^{-\alpha}$	α	α
Gaussian	$x \in [-\infty, \infty]$	$f_X(x) = \frac{1}{\sqrt{2\pi\sigma^2}} \exp\left(-\frac{(x-\mu)^2}{2\sigma^2}\right)$	μ	σ^2

to Poisson and Gaussian random variables and their associated processes, due to their preeminence in later work.

2.3.1 Poisson Statistics

Poisson random variables occur frequently in physics when counting the number of occurrences of a "rare" event of interest N. Particularly pertinent to this thesis is the example of counting the number of photons incident onto a single photon detector within a fixed time interval. To derive the probability of observing n events in the fixed time interval $[t, t+\tau]$, denoted $p_N(n; t, t+\tau)$ a number of assumptions must be made. The derivation given here closely follows that of [7] however alternative proofs can be found in [6, 18, 19]. Firstly, it must be assumed that for a very short time interval, δt, the probability of a single instantaneous event occurring is given by the product $R\delta t$ where R is called the rate i.e. $p_N(1; t, t+\delta t) = R\delta t$. Meanwhile, it is further asserted that during the interval δt the probability of two events occurring is negligible such that $p_N(0; t, t+\delta t) = 1 - R\delta t$. It is these assumptions which are responsible for what is commonly referred to as the law of rare events [10]. This law asserts that for a large number of sequential Bernoulli trials (see Table 2.1) in which the chance of success (as denoted by occurrence of 1) is low, the number of successes is a random variable approximately governed by Poisson statistics. Finally, the number of impulse events occurring in non-overlapping time intervals is assumed to be statistically independent.[2]

Consider now the probability of observing n events in the interval $t+\tau$ to $t+\tau+\delta\tau$. Since $\delta\tau$ is assumed small there are only two ways in which n events are registered: there are n occurrences in the interval $[t, t+\tau]$ and zero in $[t+\tau, t+\tau+\delta\tau]$

---

[2] If this assumption of independence between disjoint time intervals is relaxed one enters the domain of quantum light sources, which possess different photon statistics. For example photon number state sources [6] emit a single photon at very precisely defined regular times.

2.3 Some Important Types of Statistics

or alternatively there are $n - 1$ occurrences in the interval $[t, t + \tau]$ and one in $[t+\tau, t+\tau+\delta\tau]$. Hence from the assertion of statistical independence and Eq. (2.29)

$$\begin{aligned}p_N(n; t, t + \tau + \delta\tau) &= p_N(n; t, t + \tau)p_N(0; t + \tau, t + \tau + \delta\tau) \\ &\quad + p_N(n - 1; t, t + \tau)p_N(1; t + \tau, t + \tau + \delta\tau), \\ &= p_N(n; t, t + \tau)(1 - R\delta\tau) + p_N(n - 1; t, t + \tau)R\delta\tau.\end{aligned} \quad (2.42)$$

Routine algebraic manipulation then yields

$$\frac{p_N(n; t, t + \tau + \delta\tau) - p_N(n; t, t + \tau)}{\delta\tau} = R[p_N(n - 1; t, t + \tau) - p_N(n; t, t + \tau)], \quad (2.43)$$

or in the limit $\delta\tau \to 0$

$$\frac{\partial p_N(n; t, t + \tau)}{\partial \tau} = R[p_N(n - 1; t, t + \tau) - p_N(n; t, t + \tau)], \quad (2.44)$$

which holds for all n. As such Eq. (2.44) defines a family of differential equations. For $n = 0$ this reads

$$\frac{\partial p_N(0; t, t + \tau)}{\partial \tau} = -R\, p_N(0; t, t + \tau), \quad (2.45)$$

since $n \geq 0$, which, with the boundary condition $p_N(0; t, t) = 1$, can be solved immediately via standard methods (see e.g. [3]) to give $p_N(0; t, t+\tau) = \exp(-R\tau)$. Via an inductive proof it then follows that

$$p_N(n; t, t + \tau) = \frac{(R\tau)^n}{n!} \exp(-R\tau). \quad (2.46)$$

With a view to future requirement, it is useful to note a result pertaining to the sum of two (or more) independent Poisson variables. Specifically, since a Poisson random variable fundamentally arises from counting, the statistics of the sum of counting multiple variables are not changed, except the rate at which events occur increases. More mathematical rigour can be accorded to this statement [16] however this is omitted here for brevity.

Closely related to the Poisson random variable is the Poisson impulse process. Were a time trace taken of, for example, the voltage on a photomultiplier tube, a series of impulse peaks would be seen, with each possible time trace constituting a realisation. Conceivably the rate may also itself be a function of time, in which case the solution to the family of differential equations requires slight modification. Such a case was considered in for example [7, 20] in which case the resulting PDF takes the form

$$p_N(n; t, t+\tau) = \frac{(\int_t^{t+\tau} \mathcal{R}(t')dt')^n}{n!} \exp\left(-\int_t^{t+\tau} \mathcal{R}(t')dt'\right). \tag{2.47}$$

where $\mathcal{R}(t)$ denotes the instantaneous rate. The integral term can be interpreted as a total average number of events during the interval of interest and can be written in the form $\langle \mathcal{R}(t) \rangle_\tau \tau$, where $\langle \mathcal{R}(t) \rangle_\tau$ is the time averaged rate over the interval $[t, t+\tau]$. This result will be required and further developed when considering a time varying system in Chap. 8.

2.3.2 Gaussian Statistics

Whilst Poisson random variables encapsulate the statistics for rare events, it can be argued that Gaussian random variables describe the statistics when there are a large number of events. Poisson statistics arose from the sum of independent Bernouilli random variables, whereas Gaussian statistics arise from the sum of a large number of independent, identically distributed random variables,[3] a result known as the Central Limit Theorem [11]. To prove this claim,[4] let S_n be the sum of n independent, identically distributed random variables X_j with mean $E[X_j] = \mu$ and variance $\text{VAR}[X_j] = \sigma^2$. A rescaling is applied such that a new random variable Z_n is defined as

$$Z_n = \frac{S_n - n\mu}{\sigma\sqrt{n}} = \frac{1}{\sigma\sqrt{n}} \sum_{j=0}^n (X_j - \mu), \tag{2.48}$$

such that $E[Z_n] = 0$ and $\text{VAR}[Z_n] = 1$. It is hence implicitly assumed that μ and σ^2 are finite. As foreseen in Sect. 2.1.1.2 it is easier to consider the characteristic function of Z_n, viz.

$$\Phi_{Z_n}(\omega) = E\left[\exp(i\omega Z_n)\right], \tag{2.49}$$

$$= E\left[\exp\left(\frac{i\omega}{\sigma\sqrt{n}} \sum_{j=0}^n (X_j - \mu)\right)\right], \tag{2.50}$$

$$= E\left[\prod_{j=0}^n \exp\left(\frac{i\omega}{\sigma\sqrt{n}} (X_j - \mu)\right)\right]. \tag{2.51}$$

---

[3] It would therefore be expected that a Poisson random variable with a large mean would be well approximated by a Gaussian random variable, a result that is born out in practise.
[4] It should be noted that the derivation given is not strictly rigorous since it only proves convergence of the partition function $\exp[i\omega Z_n]$. The reader is directed to [17] for a fuller treatment.

2.3 Some Important Types of Statistics

Since the random variables X_j are independent, the order of multiplication and expectation can be reversed. Furthermore, for identically distributed variables the expectation is the same for all j, such that

$$\Phi_{Z_n}(\omega) = \left\{ E\left[\exp\left(i\omega\frac{1}{\sigma\sqrt{n}}(X_j - \mu)\right)\right]\right\}^n. \tag{2.52}$$

Again applying a power series expansion of the exponential term yields

$$E\left[\exp\left(\frac{i\omega}{\sigma\sqrt{n}}(X_j - \mu)\right)\right] = 1 + \frac{(i\omega)^2}{2n\sigma^2} E\left[(X - \mu)^2\right] + O\left(\frac{1}{n^{3/2}}\right), \tag{2.53}$$

where $E[X - \mu] = 0$ has also been used. Assuming the third order central moment is finite, the latter terms can be neglected for large n in comparison to the ω^2/n term. Substituting Eq. (2.53) into Eq. (2.52) and taking the limit $n \to \infty$ then gives

$$\Phi_{Z_\infty} = \lim_{n \to \infty} \Phi_{Z_n}(\omega) = \lim_{n \to \infty} \left\{1 - \frac{\omega^2}{2n}\right\}, \tag{2.54}$$

$$= \exp\left(-\frac{\omega^2}{2}\right), \tag{2.55}$$

which is the characteristic function of a Gaussian random variable with zero mean and unit variance. Via the transformation theory presented in Sect. 2.1.2 it is possible to show that the PDF of S_n is also a Gaussian random variable with mean $n\mu$ and variance $n\sigma^2$.

In view of the definition of a single Gaussian random variable it is possible to define jointly complex Gaussian random variables $\mathbf{X}$, for which the joint PDF is given by [12]

$$f_\mathbf{X}(\mathbf{x}) = \frac{1}{\pi^n |\mathbb{K}|} \exp\left(-(\mathbf{x} - \boldsymbol{\mu})^\dagger \mathbb{K}^{-1}(\mathbf{x} - \boldsymbol{\mu})\right), \tag{2.56}$$

where $E[\mathbf{X}] = \boldsymbol{\mu}$ and $\mathbb{K}^{-1}$ is the inverse of the covariance matrix.

The generality of the assumptions made in the preceding derivation accounts for the ubiquity of Gaussian random variables in nature. For the same reasons Gaussian processes, in which sample values taken of the process at different times are jointly Gaussian, are equally pervasive.

2.4 Conclusions

Throughout the course of this chapter many of the fundamental principles of probability theory have been reviewed. Emphasis has been placed on how random variables and processes can be defined and metrics to parameterise their behaviour. Although a complete description of stochastic systems is theoretically possible, it was seen to be frequently impractical. As a result approximate descriptions in terms of the first and second order moments were explored. The former describes the average behaviour of the system, and regularly proves meaningful when considering for example, noise perturbed systems, since it furnishes a description of the noise-free limit. The latter meanwhile gives a measure of the uncertainty present and hence, as will be discussed in the next chapter, the performance bounds of those same systems. Consequently, when studying more specific vectorial optical systems, such as polarisation state analysers or imaging systems, the ideas presented in this and the following chapter will be routinely required.

References

1. N. Abramson, *Information Theory and Coding* (McGraw Hill, New York, 1963)
2. R.B. Ash, C.A. Doléans-Dade, *Probability and Measure Theory* (Elsevier Academic Press, New York, 1999)
3. M.L. Boas, *Mathematical Methods in the Physical Sciences*, 2nd edn. (Wiley, New York, 1983)
4. V. Delaubert, N. Treps, C. Fabre, A. Maître, H.A. Bachor, P. Réfrégier, Quantum limits in image processing. Europhys. Lett. **81**, 44001 (2008)
5. W. Feller, *Probability Theory and its Applications* (Addison-Wesley, Reading, 1950)
6. M. Fox, *Quantum Optics: An Introduction*. Oxford Master Series in Physics (Oxford University Press, Oxford, 2006)
7. J.W. Goodman, *Statistical Optics* (Wiley, New York, 2004)
8. F. Hitoshi, A. Toshimitsu, N. Kunihiko, S. Yoshihisa, O. Takehiko, Blood flow observed by time-varying laser speckle. Opt. Lett. **10**(3), 104–106 (1985)
9. N.E. Huang, Z. Shen, S.R. Long, M.C. Wu, H.H. Shih, Q. Zheng, N.-C. Yen, C.C. Tung, H.H. Liu, The empirical mode decomposition and the Hilbert spectrum for nonlinear and non-stationary time series analysis. Proc. Math. Phys. Eng. Sci. **454**, 903–995 (1998)
10. K. Itô, *Introduction to Probability Theory* (Cambridge University Press, Cambridge, 1984)
11. J. Jacod, P. Protter, *Probability Essentials* (Springer, New York, 2004)
12. S.M. Kay, *Fundamentals of Statistical Signal Processing: Estimation Theory* (Prentice-Hall, London, 1993)
13. A. Khintchine, Korrelationstheorie der stationären stochastischen Prozesse. Math. Ann. **109**, 604–615 (1934)
14. A. Kolmogorov, *Grundbegriffe der Wahrscheinlichkeitsrechnung* (Springer, Heidelberg, 1933)
15. A. Kolmogorov, *Foundations of the Theory of Probability*, 2nd edn. (Chelsea Publishing, New York, 1956)
16. A. Leon-Garcia, *Probability and Random Processes for Electrical Engineering* (Addison-Wesley, Reading, 1994)
17. M. Loève, *Probability Theory*, 4th edn. (Springer, Heidelberg, 1978)
18. R. Loudon, *The Quantum Theory of Light*, 3rd edn. (Clarendon Press, Oxford, 2000)
19. L. Mandel, *Fluctuations of Light Beams*. Progress in Optics vol. 2 (North-Holland Publishing Co., Amsterdam, 1963)
20. L. Mandel, E. Wolf, *Optical Coherence and Quantum Optics* (Cambridge University Press, Cambridge, 1995)
21. W. Martin, P. Flandrin, Spectral analysis of nonstationary processes. IEEE Trans. Acoust. Speech. **33**, 1461–1470 (1985)
22. G. Matz, F. Hlawatsch, W. Kozek, Generalized evolutionary spectral analysis and the Weyl spectrum of nonstationary random processes. IEEE Trans. Signal Process. **45**, 1520–1534 (1997)

References

23. M.B. Priestley, Power spectral analysis of non-stationary random processes. J. Sound Vib. **6**(1), 86–97 (1967)
24. L.L. Scharf, *Statistical Signal Processing: Detection, Estimation, and Time Series Analysis* (Addison-Wesley, Reading, 1991)
25. C. E. Shannon, A mathematical theory of communication. Bell Syst. Tech. J. **27**, 379–423 and 623–656 (1948)
26. C.E. Shannon, Communication in the presence of noise. Proc. IRE **37**, 10–21 (1949)
27. N. Treps, V. Delaubert, A. Maître, J.M. Courty, C. Fabre, Quantum noise in multipixel image processing. Phys. Rev. A **71**, 013820 (2005)
28. A. van den Bos, A Cramér-Rao lower bound for complex parameters. IEEE Trans. Signal Process. **42**, 2859 (1994)
29. W. Weaver, *Some Recent Contributions to the Mathematical Theory of Communication* (University of Illinois Press, Urbana, 1949)
30. N. Wiener, Generalized harmonic analysis. Acta Math. **55**, 117–258 (1930)

Chapter 3
Information and Estimation Theory

> *A likely impossibility is always preferable to an unconvincing possibility.*
>
> Aristotle

3.1 Introduction

Shannon's seminal paper of 1948 [32] is popularly seen as the cornerstone of information theory. Arguably Shannon's paper addressed a significant problem of the age, namely the scientific quantification of the rather abstract concept of information. Despite numerous earlier works, predominantly by Nyquist, Küpfmüller, Gabor, and Hartley [15, 17, 22, 28], each fundamentally failed in a number of ways. Foremost of these deficiencies is that no rigorous treatment was given to the role of noise, although its importance was noted by both Nyquist and Hartley. Noise is an essential component to information transfer because random, and hence unpredictable, perturbations to an information carrying signal ultimately prohibit exact determination of the original signal or its implicit meaning. The probabilistic treatment of Shannon (and later contributors [30, 34, 39]) allowed the stochastic nature of noise to be considered in a concise and rigorous fashion, and thus a definition for information was born.

Shannon information (as this definition is now universally known) presents a measure of the statistical dependence of the input and output of an information channel, and hence presents a characterisation of the channel itself. By Shannon's own admission, his theories, although proving the existence of strategies by which the full capabilities of an information channel *can* be achieved, provide no means by which to determine *how* to do so. Furthermore, no regard is given as to how an observer, having received an ambiguous, noisy signal, can optimally extract the desired information; such is the domain of statistical detection, inference and estimation theory.

Optimality is patently a subjective notion and hence a multitude of metric functions, and associated estimation strategies, can be found in the literature. Frequently encountered are those of maximum entropy, minimum mean square error and maximum likelihood estimators (MLEs), to mention but a few (see for example [20] or [31]). Notably the first of these aims to select the signal which would provide the maximum, average, Shannon information (or entropy). Within estimation theory however a new definition of information emerges, namely Fisher information, which quantifies the performance of an ideal observer in estimating an original signal, given a noise corrupted version. Shannon and Fisher information share many properties, such as additivity [14], although they quantify distinct aspects of the information transmission process. Asymptotic relations between Shannon information and Fisher information have however been established [6, 19], whilst stochastic signals have also been considered in a so called Fisher–Shannon information plane [37]. Despite these relations and similarities, Shannon information will be considered no further. Whilst Shannon information has historically proven to be an invaluable tool for the evaluation and design of information channels, this thesis will predominantly focus on the readout stage necessary for any such channel, thus motivating the exclusive consideration of Fisher information due to its greater suitability for such a task. Furthermore due to its close relations to Fisher information, as will be discussed in Sect. 3.3.3, attention will also be restricted to maximum likelihood (ML) estimation strategies.

With this in mind, this chapter sets out to establish the central aspects of statistical inference which will be heavily used and quoted in the remainder of this thesis. Specifically Sect. 3.2 will formalise the concept of MLEs, detailing their construction and properties. Formulation of optimal estimators, however, requires incorporating a model for the noise present in the system of interest. In optical systems noise can be attributed to two sources which Delaubert et al. [9] refer to as technical and quantum noise. The former arises from imperfect experimental setup, for example mechanical vibrations or stray reflections, and can in principle be reduced to an arbitrarily low level by better system design. Quantum noise however, arising from the discrete nature of light and the stochastic nature of photon arrivals at the detector, can not. Quantum (or shot) noise limited operation has been reached in many applications, for example single molecule detection, optical coherence tomography (OCT) and astronomy [10, 13, 24]. As discussed in Chap. 2 the former, technical noise, can often be suitably described using Gaussian statistics, whilst the latter is more correctly modelled using Poissonian statistics. Explicit expressions for the MLE in both noise regimes will hence be given.

Estimators, being derived from random data, are also random variables. Accordingly, the notions of parameterising random behaviour presented in Chap. 2 can be employed to assess the performance of any estimator that may be developed. Particular attention will be paid to the mean and (co-) variance of an estimator. The former, which within the context of estimation theory is known as the bias of an estimator, quantifies the systematic error accrued during the inference process, whilst the

3.1 Introduction

latter quantifies the precision of the estimator, with a larger spread corresponding to a lower confidence in any estimate. Section 3.2.2 hence proceeds by discussing the mean and covariance of MLEs. Formally a lower bound on the covariance matrix of an estimator can be found, known as the Cramér-Rao lower bound (CRLB), and will be detailed in Sect. 3.3 including discussion of two informational metrics derived from Fisher's definition of information.

Sections 3.4 and 3.5 finally consider how the use of a priori information can be used to improve estimation algorithms. The first, extends the maximum likelihood principle to encapsulate constraints that may exist within a system, a simple example being a positivity constraint on optical intensities. Section 3.5 considers incorporation of probabilistic a priori information that may be held with regards to the signal source, for example, in binary communication channels it is known that either a logical 0 or 1 are transmitted with equal probability. Both strategies ideally improve the quality and precision to which parameters of interest can be estimated, as can be expressed using modified versions of the CRLB, thus also provoking a short discussion in this vein.

3.2 Maximum Likelihood Estimation

3.2.1 Definition

Many systems can be modeled as having an output dependent upon a set of N_w complex parameters w_j,[1] the values of which are of interest to an experimenter. However, due to transformations performed by the experimental setup and measurement device and the presence of noise the experimenter will obtain a set of N_x data points X_k. Having obtained their data, an experimenter sets to the task of estimating the value of the parameters **w** from **X**. To do so each possible parameter value must be evaluated and the best selected, hence requiring a suitable metric. Intuitively, one could consider how probable it is to observe a fixed **x**, for different **w** (assumed, at present, to be deterministic). Adopting an optimistic view of the experimental process, the most probable value is then used as the estimate $\hat{\mathbf{w}}$ of the parameter vector, where the hat notation is used to denote an estimator. Essentially this approach selects a system model (as parameterised by **w**) which is most likely to explain the observed data. Mathematically this can be performed by maximising the PDF $f_\mathbf{X}(\mathbf{x}|\mathbf{w})$, viewed as a function of **w** (also known as the likelihood function) with respect to **w**. Alternatively, since the logarithm is a monotonically increasing function this is equivalent to maximising the log-likelihood function, which is simply defined as the natural logarithm of $f_\mathbf{X}(\mathbf{x}|\mathbf{w})$. From Shannon's definition of information[2] the log-likelihood

---

[1] All of this work will assume that the quantities are complex. Hence **w** is assumed to be a column vector of the form $(\mathbf{u}, \mathbf{u}^*)^T$ (see e.g. [35]).

[2] Via an axiomatic approach to information, Shannon was lead to define the amount of information gained when a particular symbol x_i is received, as $-\log p_X(x_i)$, where $p_X(x_i)$ is the marginal or

function $L(\mathbf{x}, \mathbf{w})$ can be considered as a measure of the information received about $\mathbf{w}$ when $\mathbf{x}$ is observed.

Using the log-likelihood function a $N_w \times 1$ score vector can be constructed viz.[3]

$$\mathbf{s_w}(\mathbf{w}) = \left[\frac{\partial}{\partial \mathbf{w}} \ln f_{\mathbf{X}}(\mathbf{x}|\mathbf{w})\right]^\dagger, \qquad (3.1)$$

which, as the name suggests, grades different values of $\mathbf{w}$. Stationary points of the log-likelihood function are given by the solutions of the ML equations $\mathbf{s}(\hat{\mathbf{w}}) = \mathbf{0}$, where $\hat{\mathbf{w}}$ is then known as the maximum likelihood estimator (MLE). To determine the nature of the stationary point (and hence demonstrate that $\hat{\mathbf{w}}$ gives a maximum in the likelihood function) consider the Hessian matrix of the log-likelihood function given by

$$\begin{aligned}\mathbb{N} &= \left\{\frac{\partial}{\partial \mathbf{w}}\left[\frac{\partial}{\partial \mathbf{w}} \ln f_{\mathbf{X}}(\mathbf{x}|\mathbf{w})\right]^\dagger\right\}^\dagger, \\ &= \frac{1}{f_{\mathbf{X}}^2(\mathbf{x}|\mathbf{w})}\left\{f_{\mathbf{X}}(\mathbf{x}|\mathbf{w})\frac{\partial}{\partial \mathbf{w}}\left[\frac{\partial}{\partial \mathbf{w}} f_{\mathbf{X}}(\mathbf{x}|\mathbf{w})\right]^\dagger - \frac{\partial}{\partial \mathbf{w}} f_{\mathbf{X}}(\mathbf{x}|\mathbf{w})\left[\frac{\partial}{\partial \mathbf{w}} f_{\mathbf{X}}(\mathbf{x}|\mathbf{w})\right]^\dagger\right\}^\dagger, \\ &= \frac{1}{f_{\mathbf{X}}(\mathbf{x}|\mathbf{w})}\left\{\frac{\partial}{\partial \mathbf{w}}\left[\frac{\partial}{\partial \mathbf{w}} f_{\mathbf{X}}(\mathbf{x}|\mathbf{w})\right]^\dagger\right\}^\dagger - \left[\frac{\partial}{\partial \mathbf{w}} \ln f_{\mathbf{X}}(\mathbf{x}|\mathbf{w})\right]^\dagger \frac{\partial}{\partial \mathbf{w}} \ln f_{\mathbf{X}}(\mathbf{x}|\mathbf{w}). \end{aligned}$$
(3.2)

Since $\mathbb{N}$ can not be written in the form $\mathbb{A}\mathbb{A}^\dagger$, no comment can be made with regards to its definiteness [5] and hence as to the nature of the stationary point. It should be noted however that since the observed data vector is a random variable, then so too are the score vector and the estimator $\hat{\mathbf{w}}$. Average properties of $\mathbf{s_w}$ and $\hat{\mathbf{w}}$ can then be examined. In particular consider the mean of the score vector and its gradient $\mathbb{N}$;

$$\begin{aligned}E[\mathbf{s_w}(\mathbf{w})] &= E\left[\left(\frac{\partial}{\partial \mathbf{w}} \ln f_{\mathbf{X}}(\mathbf{x}|\mathbf{w})\right)^\dagger\right], \\ &= \int \left(\frac{\partial}{\partial \mathbf{w}} f_{\mathbf{X}}(\mathbf{x}|\mathbf{w})\right)^\dagger d\mathbf{x}, \\ &= \left(\frac{\partial}{\partial \mathbf{w}} \int f_{\mathbf{X}}(\mathbf{x}|\mathbf{w}) d\mathbf{x}\right)^\dagger = \mathbf{0}, \end{aligned} \qquad (3.3)$$

(Footnote 2 continued.)
a priori probability associated with receiving x_i. The base of the logarithm is normally taken as two but a different choice only corresponds to a change of units. An analogous formulation for continuous random variables exists in which information can be defined as $-\log f_X(x)$.

[3] Matrix calculus will be seen to play an important role in this work. Various authors use different conventions in this regard however those detailed in [5] are adopted henceforth.

3.2 Maximum Likelihood Estimation

where **0** is a column vector of zeros and it has been assumed that the order of differentiation and integration can be interchanged.[4] By a similar argument the expectation of first term of the Hessian matrix averages to zero, leaving

$$E[\mathbb{N}] = -E\left[\left(\frac{\partial}{\partial \mathbf{w}} \ln f_\mathbf{X}(\mathbf{x}|\mathbf{w})\right)^\dagger \frac{\partial}{\partial \mathbf{w}} \ln f_\mathbf{X}(\mathbf{x}|\mathbf{w})\right]. \qquad (3.4)$$

Since the kernel of the expectation is now of the form $\mathbb{A}\mathbb{A}^\dagger$ it is positive semi-definite, implying that the Hessian matrix is *on average* negative definite, or in other words, the solutions of $\mathbf{s}_\mathbf{w}(\hat{\mathbf{w}}) = \mathbf{0}$, on average, give maxima in the log-likelihood function. Should multiple solutions exist to the ML equations then further investigation must be made to identify the global maximum. It will however henceforth be assumed that the solution to the ML equations is unique.

3.2.2 Properties of the Maximum Likelihood Estimator

Having estimated the parameters of interest, **w**, the quality of the estimator must be judged so as to quantify the level of confidence to be held in $\hat{\mathbf{w}}$. Usually this is done using the bias $\mathbf{b}_\mathbf{w} = E[\hat{\mathbf{w}}] - \mathbf{w}$ and covariance matrix $\mathbb{K}_\mathbf{w} = E[(\hat{\mathbf{w}} - E[\hat{\mathbf{w}}])(\hat{\mathbf{w}} - E[\hat{\mathbf{w}}])^\dagger]$. These parameters will naturally differ between different noise models, therefore the performance of the MLE in Gaussian and Poisson noise regimes are hence both considered here.

3.2.2.1 Maximum Likelihood Estimation in Gaussian Noise

To determine the form and performance of the MLE in Gaussian noise (introduced in Sect. 2.3.2), it is first necessary to calculate the log-likelihood function. Substitution of Eq. (2.56) gives

$$\ln f_\mathbf{X}(\mathbf{x}|\mathbf{w}) = -n \ln(\pi|\mathbb{K}|) - (\mathbf{x} - \boldsymbol{\mu})^\dagger \mathbb{K}^{-1}(\mathbf{x} - \boldsymbol{\mu}), \qquad (3.5)$$

where the mean, $\boldsymbol{\mu} = \boldsymbol{\mu}(\mathbf{w})$, is dependent on **w**, whilst it has been assumed that the covariance matrix $\mathbb{K}$ is independent of the parameter values.

From Eqs. (3.1) and (3.5) the score vector is given by

$$\mathbf{s}_\mathbf{w} = 2\left(\frac{\partial \boldsymbol{\mu}}{\partial \mathbf{w}}\right)^\dagger \mathbb{K}^{-1}(\mathbf{x} - \boldsymbol{\mu}), \qquad (3.6)$$

---

[4] If the PDF $f_\mathbf{X}(\mathbf{x}|\mathbf{w})$ has a finite support in **x**, this assertion requires that the bounds do not depend on **w**. Alternatively, if the PDF has infinite support it requires the integral converge for all **w**. These conditions will be assumed to hold throughout this thesis.

where the fact that $\mathbb{K}$ is Hermitian has been used. Equating this equation to zero and solving for $\mathbf{w}$ yields the MLE. Such a solution cannot be expressed in closed form, thus to illustrate the properties of the MLE consider a linear model in which $\boldsymbol{\mu} = \mathbb{H}\mathbf{w}$. Within this construction the score vector reduces to

$$\mathbf{s_w} = 2\mathbb{H}^\dagger \mathbb{K}^{-1}(\mathbf{x} - \mathbb{H}\mathbf{w}), \tag{3.7}$$

such that

$$\hat{\mathbf{w}} = \left[\mathbb{H}^\dagger \mathbb{K}^{-1} \mathbb{H}\right]^{-1} \mathbb{H}^\dagger \mathbb{K}^{-1} \mathbf{x}, \tag{3.8}$$

where it is assumed the inverse $\left[\mathbb{H}^\dagger \mathbb{K}^{-1} \mathbb{H}\right]^{-1}$ exists. The bias can then be explicitly evaluated by considering

$$\begin{aligned}\mathbf{b_w} = E[\hat{\mathbf{w}} - \mathbf{w}] &= \left[\mathbb{H}^\dagger \mathbb{K}^{-1} \mathbb{H}\right]^{-1} \mathbb{H}^\dagger \mathbb{K}^{-1} E[\mathbf{x}] - \mathbf{w}, \\ &= \left[\mathbb{H}^\dagger \mathbb{K}^{-1} \mathbb{H}\right]^{-1} \mathbb{H}^\dagger \mathbb{K}^{-1} \boldsymbol{\mu} - \mathbf{w}, \\ &= \mathbf{w} - \mathbf{w} = \mathbf{0}, \end{aligned} \tag{3.9}$$

where the third equality follows from the fact that $\boldsymbol{\mu} = \mathbb{H}\mathbf{w}$ and that $\mathrm{A}^{-1}\mathrm{A} = \mathbb{I}$, where $\mathbb{I}$ is the identity matrix. The MLE in Gaussian noise is thus seen to be unbiased. Similarly the covariance matrix can be found according to

$$\begin{aligned}\mathbb{K}_\mathbf{w} &= E\left[(\hat{\mathbf{w}} - E[\hat{\mathbf{w}}])(\hat{\mathbf{w}} - E[\hat{\mathbf{w}}])^\dagger\right] = E\left[(\hat{\mathbf{w}} - \mathbf{w})(\hat{\mathbf{w}} - \mathbf{w})^\dagger\right], \\ &= E\left[\left(\left[\mathbb{H}^\dagger \mathbb{K}^{-1} \mathbb{H}\right]^{-1} \mathbb{H}^\dagger \mathbb{K}^{-1} \mathbf{x} - \mathbf{w}\right)\left(\mathbf{x}^\dagger \mathbb{K}^{-\dagger} \mathbb{H}\left[\mathbb{H}^\dagger \mathbb{K}^{-1} \mathbb{H}\right]^{-\dagger} - \mathbf{w}^\dagger\right)\right].\end{aligned} \tag{3.10}$$

Expanding the brackets yields

$$\mathbb{K}_\mathbf{w} = \left[\mathbb{H}^\dagger \mathbb{K}^{-1} \mathbb{H}\right]^{-1} \mathbb{H}^\dagger \mathbb{K}^{-1} E\left[\mathbf{x}\mathbf{x}^\dagger\right] \mathbb{K}^{-\dagger} \mathbb{H} \left[\mathbb{H}^\dagger \mathbb{K}^{-1} \mathbb{H}\right]^{-\dagger} - \mathbf{w}\mathbf{w}^\dagger. \tag{3.11}$$

Finally noting that $E[\mathbf{x}\mathbf{x}^\dagger] = \mathbb{K} + \mathbb{H}\mathbf{w}\mathbf{w}^\dagger\mathbb{H}^\dagger$, gives the $N_w \times N_w$ covariance matrix of the MLE as $\mathbb{K}_\mathbf{w} = [\mathbb{H}^\dagger \mathbb{K}^{-1} \mathbb{H}]^{-1}$.

3.2.2.2 Maximum Likelihood Estimation in Poisson Noise

Shifting attention now to the MLE in the presence of Poisson noise, a derivation of the bias and covariance, analogous to that given in [2], is given. Although not novel,

3.2 Maximum Likelihood Estimation

reproduction of this derivation is insightful, firstly to understand the conditions under which the MLE performs well, as shall be discussed later in Sect. 3.3, but also since the same logic can be applied to constrained estimation (see Sect. 3.4.1.2), which to the best of the author's knowledge has not previously been reported.

Consider then a set of N_x Poisson random variables, X_j, assumed to be independent. Such an assumption is valid, for example, when considering quantum noise infecting the readout from multiple CCD pixels. Accordingly the joint PDF can be written

$$f(\mathbf{x}|\mathbf{w}) = \prod_{j=1}^{N_x} \frac{\mu_j^{x_j}}{x_j!} \exp(-\mu_j), \qquad (3.12)$$

where μ_j denotes the jth element of the mean vector $\boldsymbol{\mu}$. Therefore the log-likelihood function is given by

$$L(\mathbf{x}, \mathbf{w}) = \sum_{j=1}^{N_x} x_j \ln \mu_j - \mu_j - \ln[x_j!]. \qquad (3.13)$$

Taking the derivative as per Eq. (3.1) yields the score vector

$$\mathbf{s_w} = \left(\frac{\partial \boldsymbol{\mu}}{\partial \mathbf{w}}\right)^\dagger \left(\mathbb{K}^{-1}\mathbf{x} - \mathbf{1}\right), \qquad (3.14)$$

where $\mathbb{K}$ is the diagonal covariance matrix $\mathrm{diag}[\mu_1, \ldots, \mu_{N_x}]$[5] and $\mathbf{1}$ is a vector of ones. Written out element-wise and again assuming a linear model, the set of N_w ML equations are

$$0 = \sum_{j=1}^{N_x} H_{jk}^* \left(\frac{x_j}{\hat{\mu}_j(\hat{\mathbf{w}})} - 1\right), \quad \text{for all } k, \qquad (3.15)$$

where H_{jk} denotes the (j, k)th element of $\mathbb{H}$ and $\hat{\mu}_j(\hat{\mathbf{w}})$ is the MLE of the jth element of the mean vector $\boldsymbol{\mu}$ (see Sect. 3.2.2.3 for a discussion on the validity of this approach). The solution of these equations however can not be expressed in closed form, but can easily be found numerically, for example by the method of scoring [26].

Calculation of the bias and covariance matrix of the MLE is however possible by letting $\delta \mathbf{w} = \hat{\mathbf{w}} - \mathbf{w}$ and taking a multivariate Taylor expansion [4] of $1/\hat{\mu}_j(\mathbf{w} + \delta \mathbf{w})$, which to first order is

---

[5] This is the appropriate covariance matrix since for a Poisson random variable the mean and variance are equal (see Table 2.1). Furthermore, the covariance matrix for independent random variables is diagonal as discussed in Sect. 2.1.3.2.

$$\frac{1}{\hat{\mu}_j(\mathbf{w}+\delta\mathbf{w})} \approx \frac{1}{\mu_j(\mathbf{w})} - \frac{1}{\mu_j^2(\mathbf{w})} \frac{\partial \mu_i(\mathbf{w})}{\partial \mathbf{w}} \cdot \delta\mathbf{w},$$

$$\approx \frac{1}{\mu_j(\mathbf{w})} - \frac{1}{\mu_j^2(\mathbf{w})} \sum_{k=0}^{N_w} H_{ik}\,\delta w_k. \qquad (3.16)$$

Substituting Eq. (3.16) into Eq. (3.15) and further letting $\mathbf{x} = \boldsymbol{\mu} + \delta\mathbf{x}$ gives

$$0 \approx \sum_{j=1}^{N_x} H_{jk}^* \left([\mu_j + \delta x_j] \left[\frac{1}{\mu_j} - \frac{1}{\mu_j^2} \sum_{k=0}^{N_w} H_{ik}\,\delta w_k \right] - 1 \right),$$

$$\approx \sum_{j=1}^{N_x} \frac{\delta x_j}{\mu_j} H_{jk}^* - \sum_{j=1}^{N_x} \sum_{k=0}^{N_w} \frac{1}{\mu_j} H_{ik} H_{jk}^* \,\delta w_k, \quad \text{for all } k, \qquad (3.17)$$

or in matrix form

$$\mathbf{0} \approx \mathbb{H}^\dagger \mathbb{K}^{-1} \delta\mathbf{x} - \mathbb{H}^\dagger \mathbb{K}^{-1} \mathbb{H}\, \delta\mathbf{w}. \qquad (3.18)$$

Assuming the inverse $[\mathbb{H}^\dagger \mathbb{K}^{-1} \mathbb{H}]^{-1}$ exists, algebraic manipulation gives

$$\delta\mathbf{w} \approx \left[\mathbb{H}^\dagger \mathbb{K}^{-1} \mathbb{H}\right]^{-1} \mathbb{H}^\dagger \mathbb{K}^{-1} \delta\mathbf{x}. \qquad (3.19)$$

The bias and covariance of the MLE to a first order approximation can then be examined by considering

$$\mathbf{b_w} \approx E[\delta\mathbf{w}] = \mathbf{0}, \qquad (3.20)$$

$$\mathbb{K_w} \approx E[\delta\mathbf{w}\delta\mathbf{w}^\dagger] = [\mathbb{H}^\dagger \mathbb{K}^{-1} \mathbb{H}]^{-1}, \qquad (3.21)$$

since $E[\delta\mathbf{x}] = \mathbf{0}$ and $E[\delta\mathbf{x}\delta\mathbf{x}^\dagger] = \mathbb{K}$.

3.2.2.3 Invariance of the MLE

For any particular setup it may be desirable to form the MLE of a function of the parameters $\mathbf{w}$. Take for instance a polarimetric example (see Chap. 5) in which the state of polarisation of light is inferred from raw intensity measurements. Intensity and polarisation can be easily related in these systems, however the question arises as to whether it is admissible to calculate the MLE of the polarisation state in terms of the MLE for the recorded intensities. To answer this question consider estimating the output of the vector valued (many-to-one) function $\boldsymbol{\theta} = \boldsymbol{\Theta}(\mathbf{w})$. Following [31], the function $\boldsymbol{\Theta}$ defines a mapping of a (possibly extended) region in the domain of $\mathbf{w}$, denoted $\mathcal{W} \subseteq \mathcal{C}^p$, to a single point in the domain $\mathcal{T} \subseteq \mathcal{C}^n$ of $\boldsymbol{\theta}$, where $\mathcal{C}^j$

3.2 Maximum Likelihood Estimation

Fig. 3.1 Schematic, following [31], of the mapping $\theta = \Theta(\mathbf{w})$, from $\mathcal{W}$ to $\mathcal{T}$

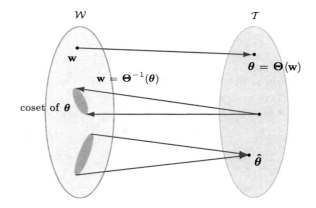

denotes the j-dimensional domain of complex numbers, as shown schematically in Fig. 3.1.

The MLE of $\mathbf{w}$ is found by maximising the log-likelihood function $L(\mathbf{x}, \mathbf{w})$, whilst the MLE of θ is found by maximising the log-likelihood function $L(\mathbf{x}, \theta)$. For a given θ, the region defined by $\Theta^{-1}(\theta)$ is known as the coset of θ. By construction the set of cosets form a disjoint covering of $\mathcal{W}$ hence implying that a given MLE $\hat{\mathbf{w}}$ lies in a single coset. Consequently it is possible to write

$$L(\mathbf{x}, \theta) = \max_{\mathbf{w} \in \Theta^{-1}(\theta)} [L(\mathbf{x}, \mathbf{w})], \qquad (3.22)$$

such that the MLE of θ is found by maximising this log-likelihood function viz.

$$\max_{\theta} [L(\mathbf{x}, \theta)] = \max_{\theta} \left[\max_{\mathbf{w} \in \Theta^{-1}(\theta)} [L(\mathbf{x}, \mathbf{w})] \right]. \qquad (3.23)$$

Equation (3.23) demonstrates that maximising either log-likelihood function with respect to the corresponding parameter vector is equivalent, such that $\hat{\theta} = \Theta(\hat{\mathbf{w}})$, i.e. the MLE of a function of a set of parameters can be found by first finding the MLE of the original parameters and using these as input into the mapping function, a result known as the invariance of the MLE.

3.3 Cramér-Rao Lower Bound

Pivotal in the theory of statistical estimation was the formulation of a lower bound on the covariance matrix of *any* estimator, known as the Cramér-Rao lower bound (CRLB) [7, 29]. The CRLB in its most commonly used form is valid for unbiased estimators only, whereby it is possible to say that the variance of *any* estimate made of the parameter vector **w** is bounded by

$$\mathbb{K}_\mathbf{w} \geq \mathbb{J}_\mathbf{w}^{-1}, \tag{3.24}$$

where the inequality implies the difference of the two matrices is positive definite, and does not necessarily hold element-wise, and $\mathbb{J}_\mathbf{w}$ is a $N_w \times N_w$ matrix known as the Fisher information matrix (FIM) [11, 12] defined by

$$\mathbb{J}_\mathbf{w} = E\left[\left(\frac{\partial \ln f_\mathbf{X}(\mathbf{x}|\mathbf{w})}{\partial \mathbf{w}}\right)^\dagger \frac{\partial \ln f_\mathbf{X}(\mathbf{x}|\mathbf{w})}{\partial \mathbf{w}}\right]. \tag{3.25}$$

Although originally derived for estimation of real parameters it was later shown that Eq. (3.24) is valid for complex parameters if the FIM is defined according to Eq. (3.25) and hence only the complex form has been given here. An estimator which achieves the CRLB, i.e. for which $\mathbb{K}_\mathbf{w} = \mathbb{J}_\mathbf{w}^{-1}$, is called efficient.

Inspection of Eqs. (3.1) and (3.25) reveals that the FIM can be expressed as the covariance matrix[6] of the score vector $\mathbb{J}_\mathbf{w} = E\left[\mathbf{s}_\mathbf{w}(\mathbf{w})\mathbf{s}_\mathbf{w}(\mathbf{w})^\dagger\right]$, that is to say it is a measure of how much the score varies as the value of **w** is varied. If all values of **w** score well then the variance of the score will be low and it is hard to choose which value of **w** is more appropriate. A large covariance or Fisher information corresponds to a stronger dependence between the observed data **X** and the parameter of interest, meaning **w** can be determined more accurately (as encapsulated by Eq. (3.24)).

Furthermore, Eq. (3.4) in conjunction with Eq. (3.25) provides a second definition for the FIM

$$\mathbb{J}_\mathbf{w} = -E\left[\left(\frac{\partial \mathbf{s}_\mathbf{w}}{\partial \mathbf{w}}\right)^\dagger\right] = -E\left[\left(\frac{\partial}{\partial \mathbf{w}}\left[\frac{\partial}{\partial \mathbf{w}}\ln f_\mathbf{X}(\mathbf{x}|\mathbf{w})\right]^\dagger\right)^\dagger\right]. \tag{3.26}$$

The second order derivative found in the kernel of this expectation can be interpreted as a measure of the curvature of the PDF as the parameter value **w** is varied. If this curvature is large this again equates to a strong dependence of the observed data **X** on **w**.

Biased estimators, however, present a further complication because they render Eq. (3.24) invalid. Bias can, for example be introduced if the parameters **w** are a non-linear function of the measured data. Under these circumstances a biased CRLB can be derived (see [18] for a derivation assuming real parameters, or a complex extension can be found in Appendix A) whereby

$$\mathbb{K}_\mathbf{w} \geq \left(\mathbb{I} + \frac{\partial \mathbf{b}_\mathbf{w}}{\partial \mathbf{w}}\right)\mathbb{J}_\mathbf{w}^{-1}\left(\mathbb{I} + \frac{\partial \mathbf{b}_\mathbf{w}}{\partial \mathbf{w}}\right)^\dagger. \tag{3.27}$$

---

[6] The score vector has zero mean as per Eq. (3.3).

3.3 Cramér-Rao Lower Bound

Efficiency of an estimator is also destroyed under the same circumstances as bias is introduced, except in the asymptotic limit of an infinite number of samples.

3.3.1 Functions of Parameters and Nuisance Parameters

As discussed previously in Sect. 3.2.2.3 it is occasionally required to estimate a function of the parameters $\mathbf{w}$, namely $\Theta(\mathbf{w})$. The invariance of the MLE allows the CRLB to be extended so as to include this scenario. Application of the chain rule when calculating the score vector gives $\mathbf{s}_\theta(\theta) = \mathbb{G}^\dagger \mathbf{s}_\mathbf{w}(\mathbf{w})$ where $\mathbb{G} = \frac{\partial \mathbf{w}}{\partial \theta}$ such that the FIM becomes

$$\mathbb{J}_\theta = \mathbb{G}^\dagger \mathbb{J}_\mathbf{w} \mathbb{G}. \qquad (3.28)$$

Hitherto it has also been assumed that full knowledge of the parameter vector $\mathbf{w}$ was desired, however this may not always be the case, an example of which is highlighted in Sect. 5.3.1.3. Assume then that the parameter vector $\mathbf{w}$ is formed via the concatenation of two vectors $\mathbf{u}$ and $\mathbf{v}$ such that $\mathbf{w} = (\mathbf{u}, \mathbf{v})$. $\mathbf{u}$ is assumed to contain the N_u desired parameters, whilst $\mathbf{v}$ contains N_v parameters that are of little interest, known as nuisance parameters. When nuisance parameters are present the relevant $N_u \times N_u$ FIM, $\mathbb{J}_\mathbf{u}$, is given by [3]

$$\mathbb{J}_\mathbf{u} = \mathbb{J}_{11} - \mathbb{J}_{12} \mathbb{J}_{22}^{-1} \mathbb{J}_{21}, \qquad (3.29)$$

where $\mathbb{J}_{11}$ is $N_u \times N_u$, $\mathbb{J}_{12}$ is $N_u \times N_v$, $\mathbb{J}_{21}$ is $N_v \times N_u$, $\mathbb{J}_{22}$ is $N_v \times N_v$ and

$$\mathbb{J}_\mathbf{w} = \begin{pmatrix} \mathbb{J}_{11} & \mathbb{J}_{12} \\ \mathbb{J}_{21} & \mathbb{J}_{22} \end{pmatrix} \qquad (3.30)$$

is the FIM for all parameters.

3.3.2 Informational Metrics

As a metric of performance, a matrix quantity, such as the FIM or estimator covariance matrix, is not ideal since its interpretation is often non-trivial. Scalar measures are thus preferable, therefore prompting the introduction of suitable quantities here. The first, known as the Fisher channel capacity $C_\mathbf{w}$ was defined by Frieden [14] and quantifies the ability of an estimation routine to extract Fisher information about the parameters of interest. Mathematically, the Fisher channel capacity is given by the trace of the FIM, i.e. $C_\mathbf{w} = \text{tr}[\mathbb{J}_\mathbf{w}]$. Noting that Eq. (3.24) also implies the weaker inequality $\text{VAR}[w_k] \geq 1/[\mathbb{J}_\mathbf{w}]_{kk}$ it can be seen that the Fisher channel capacity is a measure of the maximum total precision (as parameterised by the reciprocal of the variance) achievable. A larger $C_\mathbf{w}$ therefore describes the potential for better parameter estimation.

Alternatively one can consider the nature of the estimator $\hat{\mathbf{w}}$ itself. Since the parameter estimate is derived from random experimental data, $\hat{\mathbf{w}}$ is itself a random variable. Each experimental realisation, perturbed by differing noise, hence defines a different point in the N_w-dimensional Hilbert space in which $\mathbf{w}$ lies. For example, if attempting to estimate the Stokes vector (see Sect. 4.2.4) of polarised light from noisy intensity measurements, whereupon $\mathbf{w} = \mathbf{S}$, each estimate $\hat{\mathbf{S}}$ defines a position in Stokes space (see Sect. 4.2.6). If the estimator is unbiased, the random distribution of estimates will be centered on the true parameter value. The eigensystem of the FIM then defines the axes of a set of concentric "ellipsoids of minimum uncertainty" (in the sense of the CRLB) in Hilbert space, defined by $(\hat{\mathbf{w}} - \mathbf{w})^\dagger \mathbb{J}(\hat{\mathbf{w}} - \mathbf{w}) = c^2$. The parameter c dictates the fraction of the estimates resulting from repeated experiments which lie within the ellipsoid [31]. For example, if an unbiased efficient estimator $\hat{\mathbf{w}}$ with covariance matrix $\mathbb{J}^{-1}$ was normally distributed, then the probability that a particular estimate would lie within the region $c^2 \leq c_0^2$ could be found by integrating the N_w dimensional χ^2-squared probability distribution from 0 to c^2.

With these considerations in mind it is apparent that the volume of the so-called ellipsoids of concentration, given by

$$V_{\min} = V_{N_w}\sqrt{c^{N_w}|\mathbb{J}_\mathbf{w}^{-1}|} = V_{N_w}\sqrt{\frac{c^{N_w}}{|\mathbb{J}_\mathbf{w}|}}, \qquad (3.31)$$

can be used as a metric of aggregate error [31], where V_{N_w} is the volume of the N_w-dimensional unit hypersphere. As such the determinant of the FIM can be used as a figure of merit, whereby a larger determinant is preferable. This criterion is known as D-optimality [25, 38] and has the important advantage that it is invariant under linear operations performed on the output [25]. In the presence of nuisance parameters the appropriate matrix determinant is

$$|\mathbb{J}_\mathbf{u}| = \frac{|\mathbb{J}_\mathbf{w}|}{|\mathbb{J}_{22}|}. \qquad (3.32)$$

3.3.3 MLE Versus Other Estimators

In light of the CRLB it is possible to justify the sole consideration of the MLE over alternative estimators, such as a method of moments or minimum mean square estimator.[7] A well documented result in statistical fields is that if an efficient estimator exists then it is the MLE (see e.g. [31]), however if no such estimator exists then

---

[7] In the presence of Gaussian noise the minimum mean square estimator and MLE are equivalent [20].

3.3 Cramér-Rao Lower Bound

the MLE is both asymptotically unbiased and asymptotically efficient as the number of data points taken tends to infinity. These desirable properties are not mirrored by other estimators. By restricting attention to the MLE the absolute performance limits are hence investigated.

With regards to the noise models considered earlier it is possible to show that the FIM is given exactly (for a linear model) by $\mathbb{J}_\mathbf{w} = \mathbb{H}^\dagger \mathbb{K}^{-1} \mathbb{H}$ in both cases, where $\mathbb{K}$ is the appropriate covariance matrix. The MLE estimator in Gaussian noise is hence both unbiased and efficient. On the other hand, due to the approximations taken in Eq. (3.16) the MLE in the presence of Poisson noise is only approximately unbiased and efficient. That said, the MLE still maintains these properties asymptotically. In particular it has been shown [2] that if Poisson noise arises from the discrete nature of light incident on to a detector, then each photon can be considered as an independent sample and hence asymptotics correspond to increasing intensity on the detector. Quantitatively acceptable convergence was calculated, using the Central Limit Theorem, to require approximately 8,000 photons and will hence be achieved in most classical optical experiments. If such a criteria is not meet then the CRLB is not the strongest bound and one must resort to alternative bounds, such as the Barankin bound as discussed in [1, 21, 27]. Full exposition of these bounds will however not be given here.

3.4 Constrained Maximum Likelihood Estimation

Constraints on allowable parameter values, for example physical electromagnetic fields must satisfy Maxwell's equations, constitute a form of a priori information about the detected output. If incorporated correctly in the data processing stage this a priori information would be expected to reduce the effect of experimental errors (see for example Sect. 6.5). In this section this claim is investigated by considering the analog to the MLE, termed the constrained maximum likelihood estimator (CMLE), and its performance.

Conventional maximum likelihood estimation of a set of parameters $\mathbf{w}$ aims to maximise the log-likelihood function $L(\mathbf{x}, \mathbf{w}) = \ln f(\mathbf{x}|\mathbf{w})$. To incorporate parameter constraints it is possible to use the method of Lagrange multipliers [4], requiring slight modification of the the log-likelihood function and hence the score vector. Considering only linear equality constraints[8] on the parameters $\mathbf{w}$, expressed in the form $\mathbb{G}_\mathbf{w} \mathbf{w} = \mathbf{0}$ the modified log-likelihood function is given by $L(\mathbf{x}, \mathbf{w}) = \ln f(\mathbf{x}|\mathbf{w}) + \boldsymbol{\lambda}^\dagger \mathbb{G}_\mathbf{w} \mathbf{w}$ where $\boldsymbol{\lambda}$ is a vector of Lagrange multipliers. It is this modified log-likelihood function that is maximised in constrained maximum likelihood estimation.

For Gaussian and Poisson noise regimes the constrained score vectors are given by

---

[8] Inequality constraints are considered in [16], but were shown to leave the FIM unchanged.

$$\mathbf{s_w} = 2\mathbb{H}^\dagger \mathbb{K}_\mathbf{w}^{-1}[\mathbf{x} - \mathbb{H}\mathbf{w}] + \mathbb{G}_\mathbf{w}^\dagger \boldsymbol{\lambda}, \tag{3.33}$$

and

$$\mathbf{s_w} = \mathbb{H}^\dagger \left(\mathbb{K}_\mathbf{w}^{-1}\mathbf{x} - \mathbf{1}\right) + \mathbb{G}_\mathbf{w}^\dagger \boldsymbol{\lambda}, \tag{3.34}$$

respectively. Consider first the Gaussian case. Following the same steps as given in Sect. 3.2.2.1 and using the result $\mathbb{J}_\mathbf{w} = \mathbb{H}^\dagger \mathbb{K}^{-1}\mathbb{H}$ yields

$$\hat{\mathbf{w}} = \mathbb{J}_\mathbf{w}^{-1}\left(\mathbb{H}^\dagger \mathbb{K}^{-1}\mathbf{x} + \mathbb{G}_\mathbf{w}^\dagger \frac{\boldsymbol{\lambda}}{2}\right). \tag{3.35}$$

It is however required that the CMLE satisfy the parameter constraints and thus, by asserting $\mathbb{G}_\mathbf{w}\hat{\mathbf{w}} = \mathbf{0}$, Eq. (3.35) gives

$$\frac{\boldsymbol{\lambda}}{2} = -\left[\mathbb{G}_\mathbf{w}\mathbb{J}_\mathbf{w}^{-1}\mathbb{G}_\mathbf{w}^\dagger\right]^{-1}\mathbb{G}_\mathbf{w}\mathbb{J}_\mathbf{w}^{-1}\mathbb{H}^\dagger \mathbb{K}^{-1}\mathbf{x}. \tag{3.36}$$

Upon substitution into Eq. (3.35) the final form of the CMLE is found to be

$$\hat{\mathbf{w}} = \left(\mathbb{J}_\mathbf{w}^{-1} - \mathbb{J}_\mathbf{w}^{-1}\mathbb{G}_\mathbf{w}^\dagger\left[\mathbb{G}_\mathbf{w}\mathbb{J}^{-1}\mathbb{G}_\mathbf{w}^\dagger\right]^{-1}\mathbb{G}_\mathbf{w}\mathbb{J}_\mathbf{w}^{-1}\right)\mathbb{H}^\dagger \mathbb{K}^{-1}\mathbf{x}. \tag{3.37}$$

Equation (3.36) can also be shown to hold for quadratic constraints of the form

$$\mathbf{w}^\dagger \mathbb{G}_\mathbf{w}\mathbf{w} = 0. \tag{3.38}$$

Equation (3.37) therefore gives the CMLE in Gaussian noise under linear and quadratic equality constraints.

Similarly for constrained estimation in Poisson noise it is possible to derive the analog of Eq. (3.19). Starting from Eq. (3.34) and following the steps of Sect. 3.2.2.2 yields

$$\delta \mathbf{w} \approx \mathbb{J}_\mathbf{w}^{-1}\mathbb{H}^\dagger \mathbb{K}^{-1}\delta \mathbf{x} + \mathbb{J}_\mathbf{w}^{-1}\mathbb{G}^\dagger \boldsymbol{\lambda}. \tag{3.39}$$

Constraining the CMLE then implies $\mathbb{G}_\mathbf{w}\delta \mathbf{w} = 0$ such that

$$\boldsymbol{\lambda} \approx -\left[\mathbb{G}_\mathbf{w}\mathbb{J}_\mathbf{w}^{-1}\mathbb{G}_\mathbf{w}^\dagger\right]^{-1}\mathbb{G}_\mathbf{w}\mathbb{J}_\mathbf{w}^{-1}\mathbb{H}^\dagger \mathbb{K}^{-1}\delta \mathbf{x}. \tag{3.40}$$

Approximating Eq. (3.34) by means of the Taylor expansion discussed previously (Eq. (3.16)) and substitution of Eq. (3.40), gives the constrained ML equations that must be solved to find the CMLE in the form

3.4 Constrained Maximum Likelihood Estimation 41

$$0 \approx \left(\mathbb{I} - \mathbb{G}_\mathbf{w}^\dagger \left[\mathbb{G}_\mathbf{w} \mathbb{J}_\mathbf{w}^{-1} \mathbb{G}_\mathbf{w}^\dagger\right]^{-1} \mathbb{G}_\mathbf{w} \mathbb{J}_\mathbf{w}^{-1}\right) \left(\mathbb{H}^\dagger \mathbb{K}^{-1} \mathbf{x} - \mathbf{1}\right). \quad (3.41)$$

Equation (3.41) is not amenable to analytic solution for $\hat{\mathbf{w}}$ and hence will again require a numerical approach.

3.4.1 Constrained Cramér-Rao Lower Bound

The CRLB provides a means to characterise the best achievable performance for any unbiased estimator, by lower-bounding the covariance matrix of the estimator by the inverse of the FIM. For constrained estimation it would however be expected that better performance is obtainable than that specified by the CRLB. This expectation does hold and has been considered by previous authors. For example Gorman and Hero considered the constrained CRLB (CCRLB) when the unconstrained model has a non-singular FIM [16], as was also considered by Marzetta [23]. Stoica and Ng reformulated the derivation without the restriction of a non-singular FIM [33]. Results will be quoted from these papers as required in what follows.

For unconstrained problems the MLE is used due to its desirable asymptotic properties, as discussed in Sect. 3.3.3. Crowder has shown that these properties also hold for the CMLE [8]. The popularity of the MLE in the literature and this thesis is also, in part, due to the fact that if an efficient estimator exists, then the ML method will produce an efficient estimator. Fortunately this property also extends to the CMLE, thus further motivating the use of these estimators in this work. In this section it is thus shown that the CMLEs derived in the previous section are efficient. This result will hold exactly for estimation in Gaussian noise and to a first order approximation for Poisson noise.

3.4.1.1 Gaussian Noise

To prove the efficiency of the CMLE in Gaussian noise reference need only be made to earlier results and the literature. In, for example, [16] it was shown that the CCRLB is given by

$$\mathbb{K}_\mathbf{w} \geq \mathbb{J}_\mathbf{w}^{-1} - \mathbb{J}_\mathbf{w}^{-1} \mathbb{G}_\mathbf{w}^\dagger \left[\mathbb{G}_\mathbf{w} \mathbb{J}_\mathbf{w}^{-1} \mathbb{G}_\mathbf{w}^\dagger\right]^{-1} \mathbb{G}_\mathbf{w} \mathbb{J}_\mathbf{w}^{-1} = \mathbb{B}_\mathbf{w}. \quad (3.42)$$

It is first necessary to show that the CMLE is unbiased by taking the expectation of Eq. (3.37) yielding

$$E\left[\hat{\mathbf{w}}\right] = \mathbf{w} + E[\delta \mathbf{w}] = \mathbb{B}_\mathbf{w} \mathbb{H}^\dagger \mathbb{K}^{-1} \boldsymbol{\mu}, \quad (3.43)$$

since $E[\mathbf{x}] = \boldsymbol{\mu}$. Consequently $E[\delta\mathbf{w}] = \mathbf{0}$ as required. The covariance matrix of the CMLE then follows:

$$\begin{aligned}
\mathbb{K}_\mathbf{w} &= E\left[\delta\mathbf{w}\delta\mathbf{w}^\dagger\right], \\
&= \mathbb{B}_\mathbf{w}\mathbb{H}^\dagger\mathbb{K}^{-1} E\left[\mathbf{x}\mathbf{x}^\dagger\right] \mathbb{K}^{-\dagger}\mathbb{H}\mathbb{B}_\mathbf{w}^\dagger, \\
&= \mathbb{B}_\mathbf{w}\mathbb{H}^\dagger\mathbb{K}^{-1}\mathbb{H}\mathbb{B}_\mathbf{w}^\dagger = \mathbb{B}_\mathbf{w}.
\end{aligned} \qquad (3.44)$$

The CMLE is therefore seen to be efficient in Gaussian noise as has been previously reported in, for example, [26].

3.4.1.2 Poisson Noise

To prove the efficiency of the CMLE in Poisson noise under a first order approximation Eqs. (3.39) and (3.40) are used. Together these equations give

$$\delta\mathbf{w} \approx \mathbb{B}_\mathbf{w}\mathbb{H}^\dagger\mathbb{K}^{-1}\delta\mathbf{x}. \qquad (3.45)$$

The bias of the CMLE can be immediately examined and is again found to be zero. The covariance matrix of the CMLE is furthermore found to be equal to $\mathbb{B}_\mathbf{w}$ to within the first order approximations taken, hence demonstrating the efficiency of the CMLE in the presence of Poisson noise.

3.5 Bayesian Estimation

Equations given thus far are valid only for a particular value of $\mathbf{w}$, however the parameter values may differ between different experimental setups or measurements. The experimenter may however know from an existing model or earlier data that the object being studied belongs to a restricted class, that is to say they possess some a priori information about the parameters being measured. A fibre-optic communication channel again provides a good example, where it is known that during a measurement window either a pulse will be received or not with equal probability, representing logical 1 and 0 respectively. In these circumstances estimators are best treated using a Bayesian framework.

From the Bayesian viewpoint the parameter vector $\mathbf{w}$ is considered to be a random variable whose associated PDF, $f_\mathbf{W}(\mathbf{w})$, is known a priori (and hence known as a prior PDF). This is contrary to the classical philosophy where $\mathbf{w}$ is assumed to be deterministic and constant for multiple experiments. Accordingly it is possible to modify the definition of the FIM to accommodate this random behaviour and an estimator's a priori knowledge of its behaviour. A derivation for this scenario is given in Appendix A however the final result is restated here for completeness. Specifically

3.5 Bayesian Estimation

the Bayesian Fisher information matrix (BFIM) is given by

$$\mathbb{J}_\mathbf{w} = E_\mathbf{w}\left[\mathbb{J}_\mathbf{w}^r\right] + \mathbb{J}_\mathbf{w}^{ap}, \qquad (3.46)$$

where $\mathbb{J}_\mathbf{w}^r$ is the deterministic FIM given by Eq. (3.25) and $\mathbb{J}_\mathbf{w}^{ap}$ depends only on the prior PDF $f_\mathbf{W}(\mathbf{w})$ via

$$\mathbb{J}_\mathbf{w}^{ap} = E_\mathbf{w}\left[\left(\frac{\partial}{\partial \mathbf{w}} \ln f_\mathbf{W}(\mathbf{w})\right)^\dagger \frac{\partial}{\partial \mathbf{w}} \ln f_\mathbf{W}(\mathbf{w})\right]. \qquad (3.47)$$

A Bayesian Cramér-Rao lower bound (BCRLB) can thus be shown to hold, as was originally done by van Trees [36], whereby

$$\mathbb{K}_\mathbf{w} \geq \mathbb{J}_\mathbf{w}^{-1}, \qquad (3.48)$$

where $\mathbb{J}_\mathbf{w}$ is now the BFIM and $\mathbb{K}_\mathbf{w}$ is the covariance matrix of the estimator $\hat{\mathbf{w}}$. This bound is also commonly referred to as the "van Trees inequality" or "posterior CRLB".

The problem with the Bayesian approach is that the choice of a suitable prior distribution is often difficult to justify when modelling stochastic systems. Bias, however, is not a problem[9] (c.f. Eq. (3.24) which is only valid for unbiased estimators). Bayesian estimators are in fact often biased since they tend to give estimates which lean towards values which are known a priori to be more likely. If no a priori knowledge about the possible values of the random parameter is possessed, it can be argued that the prior PDF should be nearly flat such that any estimator formed will not cluster around any particular value. In the limit a uniform PDF over the admissible values of $\mathbf{w}$ can be used. Such a PDF is known as a non-informative PDF.

Again the question as to the existence of efficient estimators arises in the Bayesian paradigm. Fortunately it can be shown [36] that the *maximum a posteriori* (MAP) estimator is in many respects the Bayesian equivalent of the MLE. As such, if an efficient estimator exists in the Bayesian sense then it will be the MAP estimator otherwise the MAP estimator is asymptotically efficient and hence the BCRLB is achievable. It is important to note that whilst the MAP estimator performs optimally with regards to the BCRLB, it is a special case of more general Bayesian estimators, which perform optimally subject to alternative cost functions (a fuller discussion can, for example, be found in [31]).

---

[9] Formally, the assumption that the conditional bias $\mathbf{b}(\mathbf{w}) = E_{\hat{\mathbf{w}}|\mathbf{w}}[\hat{\mathbf{w}}] - \mathbf{w}$ satisfies the constraints

$$\lim_{\mathbf{w}\to\pm\infty} \mathbf{b}(\mathbf{w}) f(\hat{\mathbf{w}}|\mathbf{w}) = \mathbf{0} \qquad (3.49)$$

is required in a full derivation of Eq. (3.46) (see [36]). Such a constraint shall, however, be assumed to hold throughout this work.

3.6 Conclusions

By no means exhaustive, this chapter has set out to present a number of results in estimation theory that are crucial for later chapters. In so doing the maximum likelihood estimator has been introduced and a number of its properties considered in detail, including its invariance to transformations and asymptotic behaviour with regards to efficiency and bias. Owing to the noise sources in typical optical systems the explicit form of the MLE under Gaussian and Poisson noise regimes was given.

Most importantly to the remainder of this thesis, an alternative metric to Shannon's celebrated information measure was introduced, namely Fisher information, due to its greater suitability when considering the readout stage of an information channel. Fisher information, within the context of the CRLB, has seen particular emphasis in this chapter, since it is this bound which sets the limit to which experimental parameters can be estimated. Modification of the definition of Fisher information in the presence of nuisance parameters was also considered.

Finally the incorporation of different forms of a priori information into estimation protocols was discussed. Specifically, known constraints on a system allow a constrained maximum likelihood estimator to be constructed, which shares many optimal properties with the unconstrained version however exhibits better overall performance. Optimal properties again include the asymptotic, if not exact, achievement of the modified CCRLB. Probabilistic a priori information however lends itself towards use of a Bayesian estimator such as the MAP estimator, which again possesses optimal properties with regards to the BCRLB. All of these concepts will prove fruitful in later chapters when considering how various measurement systems can be improved in informational terms.

References

1. E.W. Barankin, Locally best unbiased estimates. Ann. Math. Stat. **20**, 477–501 (1949)
2. H.H. Barrett, J.L. Denny, R.F. Wagner, K.J. Myers, Objective assessment of image quality. II: Fisher information, Fourier crosstalk, and figures of merit for task performance. J. Opt. Soc. Am. A **12**, 834–852 (1995)
3. V. Bhapkar, C. Srinivasan, On Fisher information inequalities in the presence of nuisance parameters. Ann. Inst. Statist. Math. **46**, 593–604 (1994)
4. M.L. Boas, *Mathematical Methods in the Physical Sciences*, 2nd edn. (Wiley, New York, 1983)
5. M. Brookes, The matrix reference manual (2005), http://www.ee.ic.ac.uk/hp/staff/dmb/matrix/intro.html
6. N. Brunel, J.P. Nadal, Mutual information, Fisher information, and population coding. Neural Comput. **10**, 1731–1757 (1998)
7. H. Cramér, *Mathematical Methods of Statistics* (Princeton University Press, Princeton, 1946)
8. M. Crowder, On constrained maximum-likelihood estimation with non-iid observations. Ann. Inst. Stat. Math. **36**, 239–249 (1984)
9. V. Delaubert, N. Treps, C. Fabre, A. Maître, H.A. Bachor, P. Réfrégier, Quantum limits in image processing. Europhys. Lett. **81**, 44001 (2008)
10. A. Dubois, K. Grieve, G. Moneron, R. Lecaque, L. Vabre, C. Boccara, Ultrahigh-resolution full-field optical coherence tomography. Appl. Opt. **43**(14), 2874–2883 (2004)

References

11. R.A. Fisher, On the mathematical foundations of theoretical statistics. Philos. Trans. Roy. Soc. Lond. **222**, 309–368 (1922)
12. R.A. Fisher, Theory of statistical estimation. Proc. Camb. Phil. Soc. **22**, 700–725 (1925)
13. M.R. Foreman, S.S. Sherif, P. Török, Photon statistics in single molecule orientational imaging. Opt. Express **15**, 13597–13606 (2007)
14. B.R. Frieden, *Physics from Fisher Information: A Unification* (Cambridge University Press, Cambridge, 1998)
15. D. Gabor, Theory of communication. J. IEE **93**, 429–457 (1946)
16. J.D. Gorman, A.O. Hero, Lower bounds for parametric estimation with constraints. IEEE Trans. Inform. Theory **26**, 1285–1301 (1990)
17. R.V.L. Hartley, Transmission of information. Bell Syst. Tech. J. **7**, 535–563 (1928)
18. D. Johnson, Cramér-Rao bound. *Connexions* (2003), m11266
19. K. Kang, H. Sompolinsky, Mutual information of population codes and distance measures in probability space. Phys. Rev. Lett. **86**, 4958–4961 (2001)
20. S.M. Kay, *Fundamentals of Statistical Signal Processing: Estimation Theory* (Prentice-Hall, London, 1993)
21. M.F. Kijewski, S.P. Mueller, S.C. Moore, The Barankin bound: a model of detection with location uncertainty. Proc. SPIE **1768**, 153 (1992)
22. K. Küpfmüller, Uber einschwingvorgange in Wellen filtern. Elek. Nachrichtentech. **1**, 141–152 (1924)
23. T.L. Marzetta, A simple derivation of the constrained multiple parameter Cramer-Rao bound. IEEE Trans. Signal Process. **41**, 2247–2249 (1993)
24. B.J. Meers, Recycling in laser-interferometric gravitational-wave detectors. Phys. Rev. D **38**, 2317–2326 (1988)
25. R. Mehra, Optimal input signals for parameter estimation in dynamic systems-survey and new results. IEEE Trans. Automat. Contr. **19**, 753–768 (1974)
26. T.J. Moore, B.M. Sadler, Maximum-likelihood estimation, the Cramér-Rao bound and the method of scoring with parameter constraints. IEEE Trans. Signal Process. **56**, 895–908 (2008)
27. S.P. Mueller, C.K. Abbey, F.J. Rybicki, S.C. Moore, M.F. Kijewski, Measures of performance in nonlinear estimation tasks: prediction of estimation performance at low signal-to-noise ratio. Phys. Med. Biol. **50**, 3697–3715 (2005)
28. H. Nyquist, Certain factors affecting telegraph speed. Bell Syst. Tech. J. **3**, 324–352 (1924)
29. C. Rao, Information and the accuracy attainable in the estimation of statistical parameters. Bull. Calcutta Math. Soc. **37**, 81–89 (1945)
30. S.O. Rice, Mathematical analysis of random noise. Bell Syst. Tech. J. **23**, 46–156 (1944)
31. L.L. Scharf, *Statistical Signal Processing: Detection, Estimation, and Time Series Analysis* (Addison-Wesley, Reading, 1991)
32. C.E. Shannon, A mathematical theory of communication. Bell Syst. Tech. J. **27**, 379–423, 623–656 (1948)
33. P. Stoica, B.C. Ng, On the Cramér-Rao bound under parametric constraints. IEEE Signal Proc. Lett. **5**, 177–179 (1998)
34. W.G. Tuller, Theoretical limitations on the rate of transmission of information. Proc. IRE **37**, 468–478 (1949)
35. A. van den Bos, A Cramér-Rao lower bound for complex parameters. IEEE Trans. Signal Process. **42**, 2859 (1994)
36. H.L. van Trees, *Detection, Estimation, and Modulation Theory: Part I* (Wiley, New York, 1968)
37. C. Vignet, J.-F. Bercher, Analysis of signals in the Fisher-Shannon information plane. Phys. Lett. A **312**, 27–33 (2003)
38. E. Walter, L. Pronzatom, Qualitative and quantitative experiment design for phenomenological models—a survey. Automatica **26**, 195–213 (1990)
39. N. Weiner, *The Extrapolation, Interpolation and Smoothing of Stationary Time Series* (MIT Press, Cambridge, 1949)

Chapter 4
Vectorial Optics

> *Surely no one can find fault with the labours which eminent men have entered upon in respect of light, or into which they may enter as regards electricity and magnetism.*
>
> Michael Faraday

4.1 Electromagnetism and Optics

Maxwell's famous set of equations [52] provided a unification of a number of empirical results gathered during the 19th century, namely Gauss' flux theorems, Faraday's law of induction and Ampère's circuital law. Collectively these laws describe the properties of electromagnetic fields and their relation to charge and current distributions. In macroscopic differential form Maxwell's equations take the form

$$\nabla \cdot \mathbf{D} = \sigma_f, \tag{4.1}$$

$$\nabla \cdot \mathbf{B} = 0, \tag{4.2}$$

$$\nabla \times \mathbf{E} = -\frac{\partial \mathbf{B}}{\partial t}, \tag{4.3}$$

$$\nabla \times \mathbf{H} = \mathbf{J}_f + \frac{\partial \mathbf{D}}{\partial t}, \tag{4.4}$$

where $\mathbf{D}$, $\mathbf{E}$, $\mathbf{H}$ and $\mathbf{B}$ are the electric displacement, electric field, magnetic field and magnetic flux density respectively, whereas σ_f and $\mathbf{J}_f$ are the free charge and current densities. To provide a complete physical description of a system, Maxwell's equations must also be supplemented with constitutive relations, which describe how bulk magnetisation and polarisation of a medium are induced from the presence of a field. On a more fundamental level, the constitutive relations physically describe the strength of response of bound charge and current densities in a medium and are

M. R. Foreman, *Informational Limits in Optical Polarimetry and Vectorial Imaging*,
Springer Theses, DOI: 10.1007/978-3-642-28528-8_4,
© Springer-Verlag Berlin Heidelberg 2012

given by

$$\mathbf{D} = \epsilon \mathbf{E}, \quad (4.5)$$
$$\mathbf{B} = \mu \mathbf{H}, \quad (4.6)$$

under a harmonic bounding potential approximation [9], where ϵ and μ are the electric permittivity and magnetic permeability of a material, assumed isotropic respectively.[1] The permittivity and permeability of a medium are often expressed in relation to those of free space, $\epsilon_0 = 8.854 \times 10^{-12}$ Fm^{-1} and $\mu_0 = 4\pi \times 10^{-7}$ Hm^{-1}, such that $\epsilon = \epsilon_r \epsilon_0$ and $\mu = \mu_r \mu_0$, where ϵ_r and μ_r are then termed the relative permittivity and permeability respectively.

Upon formulation of these equations, it is possible to derive a wave equation for the electromagnetic field, which easily follows by substituting Eq. (4.4) into the curl of Eq. (4.3) in conjunction with Eqs. (4.5) and (4.6), giving

$$\nabla \times \nabla \times \mathbf{E} = \mu \left(\frac{\partial \mathbf{J}_f}{\partial t} + \epsilon \frac{\partial^2 \mathbf{E}}{\partial t^2} \right). \quad (4.7)$$

In source free regions whereby $\sigma_f = 0$ and $\mathbf{J}_f = \mathbf{0}$, the vector wave equation reduces to

$$\nabla^2 \mathbf{E} = -\epsilon \mu \omega^2 \mathbf{E}, \quad (4.8)$$

where the vector identity $\nabla \times \nabla \times \mathbf{A} = \nabla(\nabla \cdot \mathbf{A}) - \nabla^2 \mathbf{A}$ has been used [3] and an $\exp(-i\omega t)$ time dependence assumed. ω is thus the frequency of the electromagnetic wave travelling at a speed $v = 1/\sqrt{\epsilon\mu}$. Substitution of experimental values of ϵ and μ gave agreement with the speed of light, prompting Maxwell to propose light as a manifestation of electromagnetism, further supporting observations by Faraday [11] of a change in the polarisation of an optical field by application of a strong magnetic field; a magneto-optical phenomena now bearing his name.

Optics is thus merely electromagnetism specialised to frequencies of approximately 3×10^{14}–1×10^{15} Hz (or equivalently wavelengths of 300–1,000 nm). This range of frequencies not only encompasses the visible spectrum, but also extends into the near infra-red and ultra-violet regimes. Throughout the rest of this work only (quasi-)monochromatic electromagnetic fields that lie within this frequency range will be considered. A fuller discussion of Maxwell's equations and some of the numerous results that follow, can be found in any good electromagnetism textbook e.g. [32, 69].

---

[1] If the medium is not isotropic then ϵ and μ must be replaced by permittivity and permeability tensors.

4.2 Polarisation of Light

Maxwell's theory of electromagnetism is a vector field theory implying optical fields are also inherently vectorial. This property was first observed in optics (before Maxwell) by Bartholinus, a Danish mathematician studying the optical properties of calcite ($CaCO_3$) [2]. He observed that light passing through calcite could produce two displaced images, a phenomenon now called double refraction, or more commonly birefringence, however this effect was poorly understood until the works of Huygens wave theory published in 1690 [28] and Newton's corpuscular theory of 1704 [57].

Polarisation is a property that can be identified for any vector field and simply refers to the time evolution of a field vector at a fixed point in space. For light however four different field vectors can be identified, namely **E**, **D**, **B** or **H**. Due to the relationships between these different field quantities, as embodied by Maxwell's equations, it is sufficient to use only one to describe the polarisation of an electromagnetic field. The choice is arbitrary, however by convention the electric field vector **E** is used since it is the electric field that is dominant in many of the interactions between light and matter, see for example [15]. This definition will thus be used throughout this work.

By way of an illustration consider two important states of polarisation, namely linearly polarised and circularly polarised light. Linearly polarised light is light for which the electric field vector oscillates in a single plane. Linearly polarised light can be described in terms of the angle at which the plane of oscillation lies relative to a reference direction, which shall henceforth be taken as the x-axis, where the normal convention of taking the z-axis as the optical axis of a system is also adopted. In particular, the terms horizontally and vertically polarised light will be frequently used and should be taken to be light oscillating in a plane at $0°$ and $90°$ to the x-axis respectively.

Circularly polarised light possesses an electric field vector which draws out circles in a plane transverse to the direction of propagation with time. This can, for example, be produced by the superposition of two orthogonally linearly polarised fields, which oscillate $90°$ out of phase. This rotation can be in a right (clockwise) or left handed (anti-clockwise) sense (unsurprisingly termed right and left circularly polarised light respectively), where optical convention is to consider the direction of rotation when viewing the light from positive z towards the origin.

A number of different formulations exist to describe the polarisation state of light and frequent use and reference will be made of many of these during this thesis. As such, it is pertinent to discuss here the basic mathematical definitions and relations between them before any further discussion can be made. What follows is by no means intended to be a complete reference on polarised light, for many works already exist, e.g. [1, 20], but should provide the necessary results and background assumed in later chapters. A summary of different representations of some common states of polarisation is given in Table 4.1.

Table 4.1 Different representations for common states of polarisation

	Horizontal	Linear at angle ϑ	Right/Left circular	Elliptical	Unpolarised
Lissajous diagram					—
Jones vector	$E_0 e^{i\delta} \begin{pmatrix} 1 \\ 0 \end{pmatrix}$	$E_0 e^{i\delta} \begin{pmatrix} \cos\vartheta \\ \sin\vartheta \end{pmatrix}$	$\frac{E_0 e^{i\delta}}{\sqrt{2}} \begin{pmatrix} 1 \\ \pm i \end{pmatrix}$	$E_0 e^{i\delta} \begin{pmatrix} \cos\vartheta\cos\varepsilon - i\sin\vartheta\sin\varepsilon \\ \sin\vartheta\cos\varepsilon + i\cos\vartheta\sin\varepsilon \end{pmatrix}$	—
Coherency matrix	$I_0 \begin{pmatrix} 1 & 0 \\ 0 & 0 \end{pmatrix}$	$I_0 \begin{pmatrix} \cos^2\vartheta & \cos\vartheta\sin\vartheta \\ \cos\vartheta\sin\vartheta & \sin^2\vartheta \end{pmatrix}$	$\frac{I_0}{2} \begin{pmatrix} 1 & \mp i \\ \pm i & 1 \end{pmatrix}$	$\frac{I_0}{2} \begin{pmatrix} 1+\cos 2\vartheta\cos 2\varepsilon & \sin 2\vartheta\cos 2\varepsilon - i\sin 2\varepsilon \\ \sin 2\vartheta\cos 2\varepsilon + i\sin 2\varepsilon & 1-\cos 2\vartheta\cos 2\varepsilon \end{pmatrix}$	$\frac{I_0}{2} \begin{pmatrix} 1 & 0 \\ 0 & 1 \end{pmatrix}$
Stokes vector	$S_0 \begin{pmatrix} 1 \\ 1 \\ 0 \\ 0 \end{pmatrix}$	$S_0 \begin{pmatrix} 1 \\ \cos 2\vartheta \\ \sin 2\vartheta \\ 0 \end{pmatrix}$	$S_0 \begin{pmatrix} 1 \\ 0 \\ 0 \\ \pm 1 \end{pmatrix}$	$S_0 \begin{pmatrix} 1 \\ \cos 2\vartheta\cos 2\varepsilon \\ \sin 2\vartheta\cos 2\varepsilon \\ \sin 2\varepsilon \end{pmatrix}$	$S_0 \begin{pmatrix} 1 \\ 0 \\ 0 \\ 0 \end{pmatrix}$

E_0 and δ represent the amplitude and phase of a plane wave, whilst $I_0 = S_0$ represents the intensity of (potentially partially polarised) light at a single point

4.2.1 Lissajous Diagrams and the Polarisation Ellipse

Lissajous diagrams are perhaps most directly related to the definition of polarisation already introduced. In particular, if one were to trace the end point of the electric field vector as seen from a position upstream of the wave, looking towards the origin of the coordinate system along the optical axis, one would obtain a Lissajous diagram [65]. Generally such a trace possesses the form of an ellipse, as shown in Table 4.1. Mathematically an ellipse can be fully parameterised by two variables, such as its ellipticity, $\tan \varepsilon$ (where ε is termed the ellipticity angle), and the angle of the major axis relative to the x axis, ϑ, yet to fully specify a field the absolute amplitude, E_0 and phase δ of the field vector at $t = 0$ are also required. Lissajous diagrams give rise to much of the terminology surrounding states of polarisation, such as linear, circular and elliptical polarisation, however do not present a particularly powerful technique in terms of problem solving and hence alternatives must be sought.

4.2.2 Jones Vectors

Jones vectors and the associated calculus (see Sect. 4.3), introduced in a series of eight papers by Clark Jones during the 1940s and 1950s [33–40], define the polarisation of a monochromatic transverse electromagnetic plane wave by specifying the complex amplitudes of the components of the electric field perpendicular to the direction of propagation. For example, the full electric field vector of a monochromatic arbitrarily polarised (but spatially homogeneous) plane wave travelling parallel to the z-axis, at time t, is given by

$$\mathbf{E}(\mathbf{r}, t) = \begin{pmatrix} E_x \\ E_y \\ 0 \end{pmatrix} \exp[i(k_z z - \omega t)], \qquad (4.9)$$

where E_x and E_y are the x and y components of the electric field respectively and $\mathbf{r} = (x, y, z)$ is a position vector defining the point at which the field is given. Accordingly the Jones vector is

$$\mathbf{E}_0 = \begin{pmatrix} E_x \\ E_y \end{pmatrix} = E_0 \, e^{i\delta} \begin{pmatrix} \cos \vartheta \cos \varepsilon - i \sin \vartheta \sin \varepsilon \\ \sin \vartheta \cos \varepsilon + i \cos \vartheta \sin \varepsilon \end{pmatrix}. \qquad (4.10)$$

Implicit in the definition of Jones vectors is however a restriction to describing collimated beams only (i.e. those with a zero component in the direction of propagation). If a beam propagates obliquely to the optical axis of a system, it is still possible to reduce the description to a two-dimensional (2D) formalism, by working in a frame of reference in which one of the coordinate axes coincides with the direction of propagation. A restriction to collimated beams is not of consequence for the "crystal optics" systems for which Jones vectors were originally designed, however increasingly modern optical systems are combining polarisation changing elements, such as retardation plates and polarisers, with elements which change the direction of propagation, such as lenses and prisms, necessitating a full three-dimensional (3D) generalisation.

Extension of the Jones vector formalism to three dimensions can fortunately be simply achieved by specifying the complex amplitude of all three field components as opposed to only the transverse components. Maxwell's equations, and the prior knowledge that a generalised Jones vector describes a single plane wave (for which $\mathbf{E}(\mathbf{r}, t) = \mathbf{E}_0 \exp[i(\mathbf{k} \cdot \mathbf{r} - \omega t)]$) automatically fixes the propagation direction as specified by the wave vector $\mathbf{k}$, according to the equations

$$\mathbf{k} \cdot \mathbf{k}^* = k^2, \tag{4.11a}$$

$$\mathbf{k} \cdot \mathbf{E}_0 = 0, \tag{4.11b}$$

$$\mathbf{k} \times (\mathbf{k} \times \mathbf{E}_0) = -k^2 \mathbf{E}_0. \tag{4.11c}$$

A generalised Jones vector is thus defined as the complex field amplitudes of all three components of the electric field with reference to a *fixed* set of coordinate axes. It should be noted that the prefix "generalised" will be frequently omitted in this work when referring to generalised Jones vectors, firstly, because many arguments given will hold true for both 2D and 3D fields, and secondly, if this does not hold, the appropriate number of dimensions should be clear from the context.

Finally a word should be said with regards to the specification of arbitrary field distributions. Such fields can be represented as a superposition of vectorial plane waves with appropriate amplitude, phase and polarisation according to

$$\mathbf{E}(\mathbf{r}) = \int_{-\infty}^{\infty} \int_{-\infty}^{\infty} \mathbf{E}_0(\mathbf{k}) \exp(i\mathbf{k} \cdot \mathbf{r}) dk_x dk_y, \tag{4.12}$$

and is known as an angular spectrum representation [66]. Here $\mathbf{E}_0(\mathbf{k})$ is the Jones vector for the plane wave with wave vector $\mathbf{k}$. Drawing an analogy with scalar Fourier optics, the angular spectrum representation can prove a powerful modelling tool, for example a simple thin lens converts points in the back focal plane to directions in the focal plane (and vice-versa). As such the field distribution in the pupil plane is seen to define the spatial frequencies in the focal plane [21]. Similar concepts have been shown to hold in the vectorial regime, whereby one refers to the Jones pupil of a system, which has been used, for example for characterisation of polarisation aberrations in an optical system [53–55]. These concepts will prove useful when later considering high numerical aperture (NA) focusing.

4.2.3 Coherency Matrices

Realistically all light sources in nature are not completely deterministic, but instead exhibit stochastic behaviour to varying degrees. Randomness of electromagnetic waves can arise for a number of reasons, for example, radiative decay of atoms occurs randomly over a typical timescale (radiative lifetime) which in turn produces a polychromatic wave. Further examples include an extended source of atoms irregularly excited, fluctuating power supplies and propagation through turbid media. On the macroscopic level the resulting effect is to produce random fluctuations in the amplitude, phase and polarisation of an electromagnetic field with time and space.

Jones vectors express the field at a given instant of time and are hence unfortunately unable to give an adequate description of partially coherent and partially polarised light in which amplitude, phase and polarisation are not completely correlated. Statistical theories of optics (including so-called coherence theory) have therefore seen fruitful and profuse development since their inception (see e.g. [87] for a historical account on the development of coherence theory). Initially these theories were limited to scalar fields [4, 22, 50, 89], however development within the vectorial regime has spawned much literature in recent years [16, 64, 70, 88]. Although a full and rigorous treatment will not be relayed here (the reader is referred to the classical references given for such an exposition), the underlying hypothesis is that for a given time and position the electric field can be considered as a complex vector random variable. If, however, the full spatial and temporal behaviour is considered, a generalisation required to account for many salient features of stochastic fields, the electric field is regarded as a random process. Due to their relation to physically observable quantities, such as intensity or fringe visibility,[2] correlation and similar second order metrics are particularly attractive in coherence theory, thus a (mutual) coherency matrix is defined quantifying the correlation of two field vectors at the same (different) time(s) and position(s) (assuming wide-sense stationarity here and henceforth). First introduced by Wiener [82, 83], although also later reintroduced by Wolf [84], the mutual coherency matrix in its most general form can be written as

$$\mathbb{C}(\mathbf{r}_1, \mathbf{r}_2, \tau) = E[\mathbf{E}(\mathbf{r}_1, 0)\mathbf{E}^\dagger(\mathbf{r}_2, \tau)], \quad (4.13)$$

where the (i, j)th element corresponds to the cross-correlation of the ith and jth field components and τ represents the time difference between sample points.[3] For a 2D (3D) treatment of the field the coherency matrix will be 2×2 (3×3). Note

---

[2] Defined as the ratio $(I_{max} - I_{min})/(I_{max} + I_{min})$, where I_{max} and I_{min} correspond to the maximum and minimum intensity observed in a fringe pattern.

[3] Wolf claims that $\mathbb{C}(\mathbf{r}, \mathbf{r}, 0)$ should be more appropriately called a polarisation matrix [89], since the off-diagonal elements give the correlation between different field components whilst only on-diagonal elements quantify the more classical scalar coherence properties, i.e. the phase and amplitude correlations. Wolf is however inconsistent when generalising to spatial distributions, reverting back to the terminology of mutual coherency matrices. For this reason and for fear of introducing further nomenclature in a field already fraught with an over abundance of specialised terminology, coherency matrix will be used throughout this work.

that the coherency matrix is a special case of the mutual coherency matrix. Finally, it should also be noted that in the majority of coherence literature the correlation matrices are given in terms of time averages as opposed to the ensemble average of Eq. (4.13). Although this definition arose from the empirical development of coherency matrices and Stokes vectors (see Sect. 4.2.4) in which the time averaging performed by physical detectors was of importance, such a replacement is sound due to the assumption of a WSS process. As discussed in Sect. 2.2.2 wide-sense stationarity implies that second order moments are independent of time (depending on a time difference alone). Coherency matrices will henceforth be given in terms of time averages, whereby Eq. (4.13) reads

$$\mathbb{C}(\mathbf{r}_1, \mathbf{r}_2, \tau) = \langle \mathbf{E}(\mathbf{r}_1, 0)\mathbf{E}^\dagger(\mathbf{r}_2, \tau) \rangle, \tag{4.14}$$

so that notation more familiar to the reader can be adopted. Without further restrictions on the class of random process under consideration (e.g. ergodicity) this replacement is not valid for higher order moments.

Section 2.2.3 highlighted the advantages of considering random processes in the spectral domain. Adopting this practice gives rise to the CSDM (c.f. Eq. (2.41))

$$\mathbb{W}(\mathbf{r}_1, \mathbf{r}_2, \omega) = \langle \widetilde{\mathbf{E}}(\mathbf{r}_1, \omega)\widetilde{\mathbf{E}}^\dagger(\mathbf{r}_2, \omega) \rangle. \tag{4.15}$$

The CSDM is equivalent to the mutual coherency matrix if considering strictly monochromatic light. Since this assumption is made throughout this work, the ω dependence will frequently be omitted for clarity. Furthermore the notation $\mathbb{W}$ and $\mathbb{C}$ will be used interchangeably.

Closely related to coherency matrices are their vectorised forms, or coherency vectors, denoted

$$\mathbf{C} = \langle \mathbf{E} \otimes \mathbf{E}^* \rangle, \tag{4.16}$$

where $\otimes$ denotes the Kronecker product. Similarly the mutual coherency vector can be defined by $\mathbf{C}(\mathbf{r}_1, \mathbf{r}_2) = \langle \mathbf{E}(\mathbf{r}_1) \otimes \mathbf{E}^*(\mathbf{r}_2) \rangle$.

Any mathematical transformation of a coherency vector, as may be used to represent the action of an optical element, can independently change all elements. This is in contrast to transformations applied to coherency matrices. Physically this means coherency vectors are more suitable when describing light propagating through systems which can decorrelate field amplitudes and phases as shall be discussed in Sect. 4.3. This could for example arise if an incoherent scattering process is involved. Such systems are traditionally termed depolarising systems, although they can also be viewed as "decohering" systems.

4.2.4 Stokes Vectors

Despite the power of coherency matrices the elements are in general complex and hence not directly measurable. Instead a real parameterisation of a stochastic field, as developed by Stokes [67] sees greater usage. The Stokes parameters, explicitly given by

$$\begin{aligned}
S_0 &= \langle |E_x|^2 + |E_y|^2 \rangle, \\
S_1 &= \langle |E_x|^2 - |E_y|^2 \rangle &= S_0 P \cos 2\varepsilon \cos 2\vartheta, \\
S_2 &= 2\langle \mathrm{Re}[E_x E_y^*] \rangle &= S_0 P \cos 2\varepsilon \sin 2\vartheta, \\
S_3 &= 2\langle \mathrm{Im}[E_x E_y^*] \rangle &= S_0 P \sin 2\varepsilon,
\end{aligned} \quad (4.17)$$

for a 2D formulation, in turn describe the total intensity of the light beam, the difference in intensity of horizontal and vertical polarised components, the difference in intensity of components polarised at $\pm 45°$ and finally the difference in intensity of right- and left-handed circularly polarised components of the beam. P is known as the degree of polarisation [1], defined as

$$P = \frac{\text{intensity of polarised component of beam}}{\text{total intensity of beam}}, \quad (4.18)$$

$$= \frac{\sqrt{S_1^2 + S_2^2 + S_3^2}}{S_0}. \quad (4.19)$$

Energy conservation dictates that $S_0^2 \geq S_1^2 + S_2^2 + S_3^2$ i.e. $0 \leq P \leq 1$. Light for which $P = 0$ is said to be unpolarised, whilst if $P = 1$ it is completely polarised.

A Stokes vector is formed when the Stokes parameters are combined into a column vector $\mathbf{S} = (S_0, S_1, S_2, S_3)^T$. Elements of the Stokes and associated coherency vectors can be related by noting that the coherency matrix can be decomposed into a basis of Hermitian trace-orthogonal matrices $\mathbb{P}_j$, known as the Pauli spin matrices due to their prominent role in quantum mechanics [68], according to

$$\mathbb{C} = \frac{1}{2} \sum_{j=0}^{3} S_j \mathbb{P}_j, \quad (4.20)$$

where

$$\mathbb{P}_0 = \begin{pmatrix} 1 & 0 \\ 0 & 1 \end{pmatrix}, \qquad \mathbb{P}_1 = \begin{pmatrix} 1 & 0 \\ 0 & -1 \end{pmatrix}, \quad (4.21)$$

$$\mathbb{P}_2 = \begin{pmatrix} 0 & 1 \\ 1 & 0 \end{pmatrix}, \qquad \mathbb{P}_3 = \begin{pmatrix} 0 & -i \\ i & 0 \end{pmatrix}. \quad (4.22)$$

More succinctly this relation can be stated as a linear matrix equation;

$$\mathbf{S} = \mathbb{A}\mathbf{C}, \tag{4.23}$$

where

$$\mathbb{A} = \begin{pmatrix} 1 & 0 & 0 & 1 \\ 1 & 0 & 0 & -1 \\ 0 & 1 & 1 & 0 \\ 0 & i & -i & 0 \end{pmatrix}. \tag{4.24}$$

Extension of the Stokes vectors to the description of 3D fields can be easily achieved by considering the set of Gell-Mann matrices $\mathbb{G}_j$ (the 3×3 analog to the Pauli matrices);

$$\mathbb{G}_0 = \begin{pmatrix} 1 & 0 & 0 \\ 0 & 1 & 0 \\ 0 & 0 & 1 \end{pmatrix}, \quad \mathbb{G}_1 = \begin{pmatrix} 1 & 0 & 0 \\ 0 & -1 & 0 \\ 0 & 0 & 0 \end{pmatrix}, \quad \mathbb{G}_2 = \begin{pmatrix} 0 & 1 & 0 \\ 1 & 0 & 0 \\ 0 & 0 & 0 \end{pmatrix}, \tag{4.25}$$

$$\mathbb{G}_3 = \begin{pmatrix} 0 & -i & 0 \\ i & 0 & 0 \\ 0 & 0 & 0 \end{pmatrix}, \quad \mathbb{G}_4 = \begin{pmatrix} 0 & 0 & 1 \\ 0 & 0 & 0 \\ 1 & 0 & 0 \end{pmatrix}, \quad \mathbb{G}_5 = \begin{pmatrix} 0 & 0 & -i \\ 0 & 0 & 0 \\ i & 0 & 0 \end{pmatrix}, \tag{4.26}$$

$$\mathbb{G}_6 = \begin{pmatrix} 0 & 0 & 0 \\ 0 & 0 & -i \\ 0 & i & 0 \end{pmatrix}, \quad \mathbb{G}_7 = \begin{pmatrix} 0 & 0 & 0 \\ 0 & 0 & 1 \\ 0 & 1 & 0 \end{pmatrix}, \quad \mathbb{G}_8 = \begin{pmatrix} 1 & 0 & 0 \\ 0 & 1 & 0 \\ 0 & 0 & -2 \end{pmatrix}/\sqrt{3}. \tag{4.27}$$

Due to the ordering and form of the Gell-Mann matrices, the Stokes parameters S_0, S_1, S_2 and S_3 have the same physical meaning (i.e. represent components of the same polarisation) in both the 2D and 3D formulations, except for the fact that the total intensity S_0 also includes the energy contribution from the longitudinal field component E_z for the 3D case. Following the same logic as for the 2D case the generalised Stokes vector is given by Eq. (4.23) except now

$$\mathbb{A} = \begin{pmatrix} 1 & 0 & 0 & 0 & 1 & 0 & 0 & 0 & 1 \\ 1 & 0 & 0 & 0 & -1 & 0 & 0 & 0 & 0 \\ 0 & 1 & 0 & 1 & 0 & 0 & 0 & 0 & 0 \\ 0 & -i & 0 & i & 0 & 0 & 0 & 0 & 0 \\ 0 & 0 & 1 & 0 & 0 & 0 & 1 & 0 & 0 \\ 0 & 0 & -i & 0 & 0 & 0 & i & 0 & 0 \\ 0 & 0 & 0 & 0 & 0 & -i & 0 & i & 0 \\ 0 & 0 & 0 & 0 & 0 & 1 & 0 & 1 & 0 \\ \frac{1}{\sqrt{3}} & 0 & 0 & 0 & \frac{1}{\sqrt{3}} & 0 & 0 & 0 & -\frac{2}{\sqrt{3}} \end{pmatrix}. \tag{4.28}$$

4.2 Polarisation of Light 57

The topic of partial polarisation and coherence of arbitrary 3D electromagnetic fields is still under active international research and debate, with the main difficulty arising from distinguishing between fully polarised and unpolarised components of the field [19, 46, 49]. Such debate however does not prove relevant to this thesis.

4.2.5 The Poincaré Sphere

In view of Eq. (4.17) a geometric interpretation can be placed on the 2D Stokes parameters as was first introduced by Poincaré [60]. Specifically for fully polarised light the Stokes parameters define a position on the surface of a unit sphere, known as the Poincaré sphere, with spherical polar coordinates $(P, \pi/2 - 2\varepsilon, 2\vartheta)$ as shown in Fig. 4.1a. Accordingly linearly polarised states of light are located on the equator, whilst points on the north (south) pole describe right (left) circularly polarised light. The Poincaré sphere exists in a Hilbert space with coordinate axes $\{\check{S}_1, \check{S}_2, \check{S}_3\}$, which shall be referred to as *Poincaré space*, where $\check{S}_i = S_i/S_0$ denotes the normalised Stokes parameters[4] for $i = 1, 2, 3$. Partially polarised states of light ($P < 1$) are described by points lying within the sphere, with unpolarised light ultimately lying at the center of the sphere [1, 7]. The Poincaré sphere has analogs in different fields of physics, such as the Ewald sphere in scattering theory and the Bloch sphere in quantum mechanics [68] and all have proven to be powerful analysis tools.

4.2.6 Stokes Space

Difficulties can arise when considering the Poincaré sphere representation of polarised light since all four degrees of freedom in **S** are not suitably depicted. When describing general partially polarised states it is perhaps more appropriate to consider the Stokes vector as defining a position in a Hilbert space $\mathcal{S}$ with coordinate axes $\{S_0, \check{S}_1, \check{S}_2, \check{S}_3\}$, which shall be referred to as *Stokes space*. Physically allowable Stokes vectors must satisfy the inequality $S_0^2 \geq S_1^2 + S_2^2 + S_3^2$ as described previously. Physical Stokes vectors therefore define a hyper-cylinder in Stokes space of unit radius, with the intensity, S_0, defining the axis of the cylinder (see Fig. 4.1b). A similar idea was proposed by Tyo [77] in which the coordinate axes were instead defined by $\{S_0, S_1, S_2, S_3\}$ meaning physical Stokes vectors spanned a *Stokes cone*. The advantage however of using normalised coordinates is that this produces a Hilbert space in which polarisation properties and intensity are orthogonal.

A slice in Stokes space perpendicular to the S_0 axis yields the Poincaré sphere, whilst a cross-section taken perpendicular to the $\check{S}_3$ axis (and through the origin) produces a hyper-cylinder with unit radius akin to that shown in Fig. 4.1b. Points

---

[4] The caron notation has been used in contrast to the conventional hat notation, so as to avoid confusion with that used for statistical estimators.

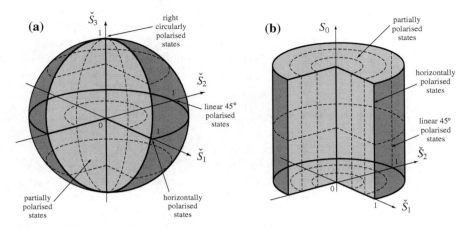

Fig. 4.1 Schematic of the (**a**) Poincaré sphere and (**b**) Stokes cylinder construction illustrated for linearly polarised light ($\check{S}_3 = 0$). The axis of the cylinder defines the intensity, S_0, axis. Totally polarised states lie on the outer surface of both the Poincaré sphere and Stokes cylinder (*dark shading*), with partially polarised states lying inside (*lighter shading*)

on the surface of this Stokes cylinder describe totally linearly polarised light, whilst points towards the central axis describe partially polarised states. Similar cross-sections taken perpendicular to $\check{S}_1$ and $\check{S}_2$ also give hyper-cylinders.

Extension of both the Poincaré sphere and Stokes cylinder representation to the description of 3D fields is fundamentally possible by increasing the dimensions of the associated Hilbert spaces from 3 (4) to 8 (9), respectively.

4.3 Jones and Mueller Polarisation Algebras

The capacity to describe (partially) polarised light serves little purpose without the further ability to propagate it through polarisation sensitive optical systems. Propagation of polarised light through non-depolarising optical systems, by means of what is now known as a Jones matrix, was addressed by Jones in his original set of papers, whilst Mueller later introduced a further matrix formulation to describe the propagation through depolarising systems [56, 59]. Initially considering only the propagation of beam like fields (i.e. a 2D treatment) through a single polarisation element[5] the former transforms Jones vectors using 2×2 matrices, $\mathbb{T}$, whilst the latter transforms Stokes vectors by means of 4×4 Mueller matrices, $\mathbb{M}$, according to

---

[5] An optical element which alters the state of polarisation of the incident light and/or is sensitive to the incident polarisation state.

4.3 Jones and Mueller Polarisation Algebras

$$\mathbf{E}_o = \mathbb{T}\mathbf{E}_i, \quad (4.29)$$

$$\mathbf{S}_o = \mathbb{M}\mathbf{S}_i, \quad (4.30)$$

where the subscripts i and o denote input and output states. Substituting Eq. (4.29) into the definition of a coherency matrix (Eq. (4.14)) it can be shown that Jones matrices can also be used to propagate coherency matrices through an optical system according to

$$\mathbb{C}_o = \mathbb{T}\mathbb{C}_i\mathbb{T}^\dagger. \quad (4.31)$$

For non-depolarising systems Eqs. (4.16), (4.23) and (4.29) give a relation between the corresponding Jones and Mueller matrices viz.

$$\mathbb{M} = \mathbb{A}(\mathbb{T} \otimes \mathbb{T}^*)\mathbb{A}^{-1}. \quad (4.32)$$

It should however be remembered that depolarising systems cannot be described using a deterministic Jones matrix and hence Eq. (4.32) does not hold in this case, since $\mathbb{T}$ does not exist. Random Jones matrices can describe depolarising systems, however the relation to the associated Mueller matrix is more complex [43]. Restriction to deterministic Jones matrices will therefore be made throughout this thesis. The Jones and Mueller matrices for some common optical elements are given in Table 4.2 for reference.

The generalisation of Jones matrices to describe 3D fields was first done heuristically by Török et al. [75, 76] in 1995. They later further extended their work [27] to include the polarisation altering characteristics of high aperture lenses first described by Inoue and Kubota [31, 44] and to describe conventional and confocal polarised light microscopes [72, 74], which will be discussed in Chap. 6.

Following [71], it is possible to define three types of generalised Jones matrices describing field transformations, interactions with surfaces, and the action of polarisation elements, such as retarders and lenses. Since a Jones vector describes a plane wave it is legitimate to associate with it an optical ray describing the direction of propagation. In many situations, such as focusing or transmission through a prism, it is necessary to describe a change in the direction of propagation of a ray of light. Generally, rotation of a ray is about a direction perpendicular to the meridional plane[6] in which the ray lies. It is thus the field components lying in the meridional plane which are rotated, whilst the perpendicular component is left unaffected. This can be simply described by the rotation matrix

$$\mathbb{L} = \begin{pmatrix} \cos\Delta\theta & 0 & \sin\Delta\theta \\ 0 & 1 & 0 \\ -\sin\Delta\theta & 0 & \cos\Delta\theta \end{pmatrix}, \quad (4.33)$$

---

[6] The plane containing the ray and the optical axis of the system.

Table 4.2 Jones and Mueller matrices for some common optical components

Optical component	Jones matrix	Mueller matrix	Notes
Linear retarder	$\begin{pmatrix} \cos\frac{\delta}{2}+i\cos 2\gamma \sin\frac{\delta}{2} & i\sin 2\gamma \sin\frac{\delta}{2} \\ i\sin 2\gamma \sin\frac{\delta}{2} & \cos\frac{\delta}{2}-i\cos 2\gamma \sin\frac{\delta}{2} \end{pmatrix}$	$\begin{pmatrix} 1 & 0 & 0 & 0 \\ 0 & 1 & 0 & 0 \\ 0 & 0 & \cos\delta & \sin\delta \\ 0 & 0 & -\sin\delta & \cos\delta \end{pmatrix}$	γ is the angle of the fast axis relative to the positive x axis. δ is the retardation introduced
Linear polariser	$\begin{pmatrix} \cos^2\gamma & \sin\gamma\cos\gamma \\ \sin\gamma\cos\gamma & \sin^2\gamma \end{pmatrix}$	$\frac{1}{2}\begin{pmatrix} 1 & \cos 2\gamma & \sin 2\gamma & 0 \\ \cos 2\gamma & \cos^2 2\gamma & \sin 2\gamma \cos 2\gamma & 0 \\ \sin 2\gamma & \sin 2\gamma \cos 2\gamma & \sin^2 2\gamma & 0 \\ 0 & 0 & 0 & 0 \end{pmatrix}$	γ is the angle of the transmission axis of the polariser relative to the positive x axis
Rotator	$\begin{pmatrix} \cos\alpha & -\sin\alpha \\ \sin\alpha & \cos\alpha \end{pmatrix}$	$\begin{pmatrix} 1 & 0 & 0 & 0 \\ 0 & \cos 2\alpha & \sin 2\alpha & 0 \\ 0 & -\sin 2\alpha & \cos 2\alpha & 0 \\ 0 & 0 & 0 & 1 \end{pmatrix}$	α is the angle by which the major axis of the polarisation ellipse is rotated
Depolariser	—	$\begin{pmatrix} 1 & 0 & 0 & 0 \\ 0 & a & 0 & 0 \\ 0 & 0 & b & 0 \\ 0 & 0 & 0 & c \end{pmatrix}$	Principal axes of the depolariser correspond to the coordinate axes. $|a|, |b|, |c| \leq 1$

4.3 Jones and Mueller Polarisation Algebras

where $\Delta\theta$ is the angle by which the ray is rotated. This matrix can account for a direction change due to refraction and reflection for example, however if the change in direction occurs at a surface there will also be a change in amplitude of the perpendicular, s, and parallel, p, components as given by the Fresnel transmission and reflection coefficients [4]. This behaviour can be described using the matrices

$$\mathbb{I}_t = \begin{pmatrix} t_p & 0 & 0 \\ 0 & t_s & 0 \\ 0 & 0 & t_p \end{pmatrix}, \qquad \mathbb{I}_r = \begin{pmatrix} r_p & 0 & 0 \\ 0 & r_s & 0 \\ 0 & 0 & r_p \end{pmatrix}. \qquad (4.34)$$

For more complex interfaces, such as stratified media, it is also possible to find similar matrices, see for example [74]. The matrices $\mathbb{L}$, $\mathbb{I}_t$ and $\mathbb{I}_r$ can only be applied when the electric field is decomposed into its s and p components. This field transformation requires a simple rotation of the Cartesian coordinate axes about the optic axis as expressed by the rotation matrix

$$\mathbb{R} = \begin{pmatrix} \cos\phi & \sin\phi & 0 \\ -\sin\phi & \cos\phi & 0 \\ 0 & 0 & 1 \end{pmatrix}, \qquad (4.35)$$

where ϕ is the angle of the meridional plane to the positive x-axis.

Coming finally to the transformations induced by polarisation elements, if the entrance and exit surfaces of the component are parallel (although possibly oblique to the incident ray vector) it is assumed, due to experimental experience, that the direction of propagation of the ray is unaltered except for a potential lateral shift which is irrelevant for a plane wave. Consequently the longitudinal field component must be unchanged upon propagation through such a parallel surface component. Hence the generalised Jones matrix for an ideal linear retarder (Babinet-Soleil compensator) is

$$\mathbb{BS} = \begin{pmatrix} \cos\frac{\delta}{2} + i\cos 2\gamma \sin\frac{\delta}{2} & i\sin 2\gamma \sin\frac{\delta}{2} & 0 \\ i\sin 2\gamma \sin\frac{\delta}{2} & \cos\frac{\delta}{2} - i\cos 2\gamma \sin\frac{\delta}{2} & 0 \\ 0 & 0 & 1 \end{pmatrix}, \qquad (4.36)$$

where γ is the azimuth of the fast axis of the retarder and δ is the relative retardation between the E_x and E_y component, whilst that for a linear polariser is

$$\mathbb{LP} = \begin{pmatrix} \cos^2\gamma & \sin\gamma\cos\gamma & 0 \\ \sin\gamma\cos\gamma & \sin^2\gamma & 0 \\ 0 & 0 & 1 \end{pmatrix}. \qquad (4.37)$$

4.4 Vectorial Ray-Tracing

Practically all optical systems are composed of multiple elements and surfaces through which light propagates. Analysis of such systems is a significant problem in optical design. Traditionally geometrical, or ray, techniques would be used in which arbitrary ray paths through an optical system are traced so as to calculate the aberrations present (or any other desired system analysis), for example by means of ABCD matrices [5]. With the increasing popularity of surface coatings on many components, thin film calculations became an equally important part of the design process to incorporate the polarisation transmission properties of these coatings [5]. Although this combined approach proves adequate for many systems in which propagation of light can be considered from a scalar viewpoint, the increasing requirement for accurate vectorial modelling of systems, e.g. in high NA imaging, a more advanced strategy is necessary. Vectorial ray tracing provides a solution to this problem [53, 79], which, by allowing a ray to represent a vectorial plane wave, can simultaneous account for both geometrical and polarisation effects. Evidently the generalised Jones formalism describes the propagation of light in exactly this way.

Complex optical systems, formed from multiple optical components can be modelled by successive application of generalised Jones matrices, i.e. for a system comprised of N sequential elements with associated Jones matrix $\mathbb{T}_j$ the whole system can be described by the composite Jones matrix

$$\mathbb{T} = \mathbb{T}_N \mathbb{T}_{N-1} \cdots \mathbb{T}_2 \mathbb{T}_1. \qquad (4.38)$$

Vectorial ray tracing through optical components using Jones matrices is advantageous over using Fresnel's equations at each individual surface of an optical system (a method used in [25] for example) since it determines an effective *single* interface which represents the entire action of the system. In this way the resulting expressions are simpler.

Vectorial ray tracing in many circumstances is unfortunately only an approximate method of modelling optical systems, in much the same way as geometrical optics cannot accurately model many scalar systems. Nevertheless it has proven of particular commercial importance having spawned a number of software packages, e.g. [8]. Furthermore under a number of conditions, which shall be discussed more fully in the context of focusing of light in the next section, vectorial ray tracing can be a rigorous calculation tool.

4.5 Focusing of Vectorial Beams

Focusing in optical systems has been researched in earnest for a myriad of different scenarios due to the pivotal role lenses play in modern optical setups. The calculation of the field distribution in the focal region of a high NA lens is an important problem

4.5 Focusing of Vectorial Beams

because of the many applications that use tightly focused beams, such as optical microscopy, optical data storage and lithography. Focusing of coherent light under a scalar approximation has been well understood for many years (see for example [66]), however when the NA of a lens exceeds approximately 0.5 the field distribution in the focal region, polarisation properties of light play a more important role necessitating the use of a full electromagnetic diffraction theory. Even as early as 1919 focusing of coherent, fully polarised electromagnetic waves could be described [29], by what is now known as the Debye-Wolf diffraction integral [62, 85] or modified Fresnel-Kirchhoff diffraction integrals [51]. In more recent years attention has slowly turned towards focusing of partially coherent light in both scalar [12, 47, 63, 80] and vectorial [46, 90] regimes due to its potential use in lithography, laser fusion and microscopy [18, 42, 45, 78]. Consideration of the full electromagnetic problem has however been limited to spatially *homogeneous* (partial) polarisation across the pupil of the focusing lens. The full and general treatment of the focusing of *inhomogeneous*, partially coherent, partially polarised waves, however, was not addressed until the work of the author [14], the details of which will be given in the following pages.

4.5.1 Scaled Debye-Wolf Diffraction Integral

Before focusing of spatially inhomogeneous partially coherent, partially polarised light can be discussed, it is necessary to first understand the principles of focusing fully coherent, totally polarised beams (although still potentially spatially inhomogeneous). For the sake of generality, systems of arbitrary Fresnel number[7] are discussed, however it is also important to consider systems of high NA, since polarisation effects decrease with NA as has been demonstrated by several authors [17, 73]. Focusing of quasi-monochromatic light in such systems can be described using the so-called scaled Debye-Wolf diffraction integral which, assuming Kirchhoff's boundary conditions at a circular aperture on an infinite screen, can be derived directly from Maxwell's equations [71] and is given by

$$\mathbf{E}(\rho) = -\frac{if^2 \exp(ik\Phi_0)}{\lambda(f+z)} \iint_{s_x^2+s_y^2 \leq 1} \mathbf{e}(s_x, s_y) \exp(ik\mathbf{s} \cdot \mathbf{P}) \frac{ds_x ds_y}{s_z}, \qquad (4.40)$$

where $\mathbf{P} = (R\cos\varphi, R\sin\varphi, Z)$ represents the position vector $\rho = (\rho\cos\varphi, \rho\sin\varphi, z)$ of a point of observation in the focal region, in a transformed co-ordinate system where

---

[7] The Fresnel number of a focusing optical system with circular aperture is defined as

$$N_F = \left(\frac{a}{f}\right)^2 \frac{f}{\lambda}, \qquad (4.39)$$

where a is the radius of the aperture, f is the focal length and λ is the wavelength of the incident light.

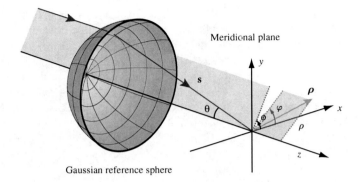

Fig. 4.2 Coordinate system and geometry of the scaled Debye-Wolf diffraction integral

$$R = \frac{f}{f+z}\rho, \tag{4.41}$$

$$Z = \frac{f}{f+z}z, \tag{4.42}$$

$\mathbf{s} = (s_x, s_y, s_z) = (\sin\theta\cos\phi, \sin\theta\sin\phi, \cos\theta)$ is a unit vector describing the direction of a ray (see Fig. 4.2), f is the focal length of the lens, $k = 2\pi/\lambda = \omega/c$ is the wavenumber of light of wavelength λ and frequency ω,

$$\Phi_0 = f + z + \frac{\rho^2 - 2fz}{2(f+z)}, \tag{4.43}$$

and $\mathbf{e}(s_x, s_y)$ describes the field distribution on the Gaussian reference sphere located in the exit pupil of the system centered on the geometrical focus of the lens. Implicit in Eq. (4.40) is the assumption that the point of observation is far from the pupil plane, such that evanescent waves can be neglected, hence yielding the stated domain of integration. Noting that an element of solid angle over the reference sphere is given by

$$\frac{ds_x ds_y}{s_z} = \sin\theta d\theta d\phi, \tag{4.44}$$

the scaled Debye-Wolf integral can be rewritten

$$\mathbf{E}(\rho, \varphi, z) = -\frac{if^2 \exp(ik\Phi_0)}{\lambda(f+z)}$$
$$\times \int_0^{2\pi}\int_0^{\alpha} \mathbf{e}(\theta, \phi) \exp[ikR\sin\theta\cos(\phi - \varphi)] e^{ikZ\cos\theta} \sin\theta d\theta d\phi \tag{4.45}$$

4.5 Focusing of Vectorial Beams

where α is the semi-angle of convergence of the lens, such that the NA (assuming the lens is in air) is given by $\text{NA} = \sin \alpha$.

4.5.1.1 High Fresnel Number Optical Systems

Microscope objective lenses are commonly designed to be telecentric, since this helps minimise aberrations and to produce a shift invariant point spread function. Consequently, the Fresnel number is also frequently large. Systems of large Fresnel numbers (whereby $R \approx \rho$ and $Z \approx z$) allow simplification of the scaled Debye-Wolf integral to

$$\mathbf{E}(\rho, \varphi, z) = -\frac{if}{\lambda} \int_0^{2\pi} \int_0^{\alpha} \mathbf{e}(\theta, \phi) \exp\left[ik\rho \sin\theta \cos(\phi - \varphi)\right] e^{ikz\cos\theta} \sin\theta \, d\theta \, d\phi. \tag{4.46}$$

Equation (4.46) is known as the Debye-Wolf integral. An equivalent equation also holds for the magnetic field vector. Equation (4.46) is also valid for systems of arbitrary Fresnel number if $z \ll f$.

Interestingly, since the lens is assumed to be telecentric from the image side, the exit pupil is located at infinity and hence $\mathbf{e}(\theta, \phi)$ must also be specified here. Specification of the field at infinity is equivalent to specifying the field in the back focal plane of the lens for a wave with infinitely small wavelength, that is to say it is only necessary to specify the field at this plane as predicted by geometrical optics as can be found using vectorial ray tracing. This is precisely the Jones pupil described in Sect. 4.2.2.

Consider then the specification of the geometric field distribution $\mathbf{e}(\theta, \phi)$ on the Gaussian reference sphere as calculated from the Jones pupil of an ideal focusing system as represented by the spatially dependent Jones vector $\widetilde{\mathbf{E}}(\theta, \phi)$. An ideal lens acts to rotate a lens by an angle $\Delta\theta = \theta$ (see Fig. 4.2) about an axis perpendicular to the meridional plane, as can be described by the generalised Jones matrix $\mathbb{L}(\theta)$ (c.f. Eq. (4.33)). It is however necessary to first decompose the field into its constituent s and p components using the generalised Jones matrix $\mathbb{R}(\phi)$. Accordingly, the field on the reference sphere, as seen from the image side of the lens, is given by (neglecting skew rays)

$$\mathbf{e}(\theta, \phi) = a(\theta) \mathbb{R}^{-1}(\phi) \cdot \mathbb{L}(\theta) \cdot \mathbb{R}(\phi) \cdot \widetilde{\mathbf{E}}(\theta, \phi) = \mathbb{Q}(\theta, \phi) \cdot \widetilde{\mathbf{E}}(\theta, \phi), \tag{4.47}$$

where $\mathbb{Q}(\theta, \phi) = a(\theta) \mathbb{R}^{-1}(\phi) \cdot \mathbb{L}(\theta) \cdot \mathbb{R}(\phi)$ and the final rotation $\mathbb{R}^{-1}$ is to transform from the s and p frame of reference back to the initial Cartesian frame. The scalar factor $a(\theta)$ is an apodisation factor that ensures energy is conserved when projecting from a plane to a sphere. For example $a(\theta) = \sqrt{\cos\theta}$ or $a(\theta) = 1$ if the lens satisfies the sine or Herschel condition respectively [30]. The Debye-Wolf integral is thus seen to represent the field in the focal region of a lens as the superposition of vectorial plane waves of appropriate strength and polarisation.

4.5.1.2 Examples

Through the Debye-Wolf integral it is possible to focus inhomogeneous fully coherent, polarised beams. A number of examples are given here to highlight the characteristics of high NA focusing. An aplanatic lens[8] will be assumed throughout. The first case discussed here is perhaps the most commonly considered in the literature, namely that of a uniformly linearly polarised beam described by the Jones vector $\tilde{\mathbf{E}} = (\mathcal{E}_x, \mathcal{E}_y, 0)$. Expanding $\mathbb{Q}(\theta, \phi)$ gives

$$\mathbb{Q}(\theta, \phi) = \frac{a(\theta)}{2} \begin{pmatrix} q_1 + q_2 \cos 2\phi & q_2 \sin 2\phi & q_3 \cos \phi \\ q_2 \sin 2\phi & q_1 - q_2 \cos 2\phi & q_3 \sin \phi \\ -q_3 \cos \phi & -q_3 \sin \phi & q_4 \end{pmatrix}, \quad (4.48)$$

where

$$q_1 = \cos \theta + 1, \quad (4.49a)$$
$$q_2 = \cos \theta - 1, \quad (4.49b)$$
$$q_3 = 2 \sin \theta, \quad (4.49c)$$
$$q_4 = 2 \cos \theta, \quad (4.49d)$$

which in turn yields

$$\mathbf{e}(\theta, \phi) = \begin{pmatrix} \mathcal{E}_x(q_1 + q_2 \cos 2\phi) + q_2 \mathcal{E}_y \sin 2\phi \\ \mathcal{E}_y(q_1 - q_2 \cos 2\phi) + q_2 \mathcal{E}_y \sin 2\phi \\ -q_3(\mathcal{E}_x \cos \phi + \mathcal{E}_y \sin \phi) \end{pmatrix}. \quad (4.50)$$

Using the well-known identity [81]

$$\int_0^{2\pi} \begin{Bmatrix} \sin m\alpha \\ \cos m\alpha \end{Bmatrix} \exp\left[ia \cos(\alpha - \beta)\right] d\alpha = 2\pi i^m \begin{Bmatrix} \sin m\beta \\ \cos m\beta \end{Bmatrix} J_m(a), \quad (4.51)$$

where $J_m(a)$ is the Bessel function of the first kind of order m, allows the azimuthal integration to be performed analytically yielding

$$\mathbf{E}(\rho, \varphi, z) = -ikf \begin{pmatrix} \mathcal{E}_x(I_0 + I_2 \cos 2\varphi) + \mathcal{E}_y I_2 \sin 2\varphi \\ \mathcal{E}_y(I_0 - I_2 \cos 2\varphi) + \mathcal{E}_x I_2 \sin 2\varphi \\ -2i I_1(\mathcal{E}_x \cos \varphi + \mathcal{E}_y \sin \varphi) \end{pmatrix}, \quad (4.52)$$

---

[8] One satisfying the sine condition.

4.5 Focusing of Vectorial Beams

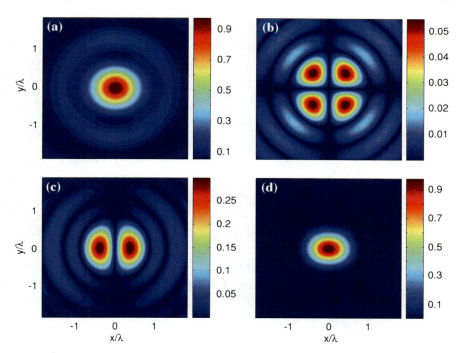

Fig. 4.3 **a**–**c** Absolute magnitude of the Cartesian field components (E_x, E_y, E_z) and **d** optical intensity in the focal plane of a lens of NA = 0.95 for x-polarised illumination

where

$$I_0 = \int_0^\alpha \sqrt{\cos\theta} \sin\theta (\cos\theta + 1) J_0(k\rho \sin\theta) \exp(ikz\cos\theta) d\theta, \quad (4.53)$$

$$I_1 = \int_0^\alpha \sqrt{\cos\theta} \sin^2\theta J_1(k\rho \sin\theta) \exp(ikz\cos\theta) d\theta, \quad (4.54)$$

$$I_2 = \int_0^\alpha \sqrt{\cos\theta} \sin\theta (\cos\theta - 1) J_2(k\rho \sin\theta) \exp(ikz\cos\theta) d\theta. \quad (4.55)$$

The focused field distribution for an x-polarised incident beam is shown in Fig. 4.3 from which it is seen a non-zero y and z component of polarisation is generated. Experimental verification of these results has been obtained for the transverse field components [61]. Determination of the longitudinal field component is more problematic, however a new technique developed by the author is detailed in Chap. 8.

Figures 4.4 and 4.5 show the focused field distribution for radially and azimuthally polarised light for which the incident Jones pupils are given by

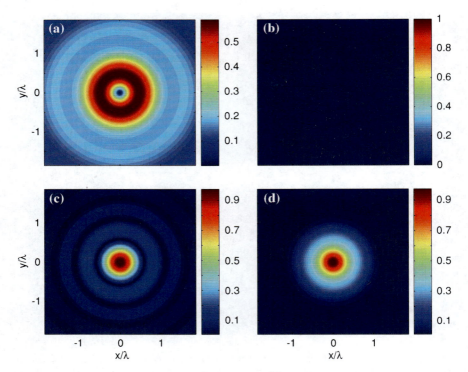

Fig. 4.4 a–c Absolute magnitude of the cylindrical field components (E_ρ, E_φ, E_z) and **d** optical intensity in the focal plane of a lens of NA = 0.95 for radially polarised illumination. Whilst E_φ is identically zero it has been shown for ease of comparison with Fig. 4.5

$$\widetilde{\mathbf{E}}_{\mathrm{rad}}(\theta,\phi) = \mathcal{E}_0 \begin{pmatrix} \cos\phi \\ \sin\phi \\ 0 \end{pmatrix}, \quad \widetilde{\mathbf{E}}_{\mathrm{azi}}(\theta,\phi) = \mathcal{E}_0 \begin{pmatrix} \sin\phi \\ -\cos\phi \\ 0 \end{pmatrix}. \tag{4.56}$$

Radially polarised beams are becoming increasingly popular because upon focusing they give a focal spot narrower than the Rayleigh diffraction limit [10]. Azimuthally polarised beams are equally seeing attention in the literature since upon focusing they produce a focal ring useful, for example, in stimulated emission depletion (STED) microscopy [26, 73]. Semi-analytic answers, similar to those given for linearly polarised input, can be found but will not be given here because more general expressions are derived in Sect. 4.5.3 for a partially coherent, partially polarised scenario.

4.5 Focusing of Vectorial Beams

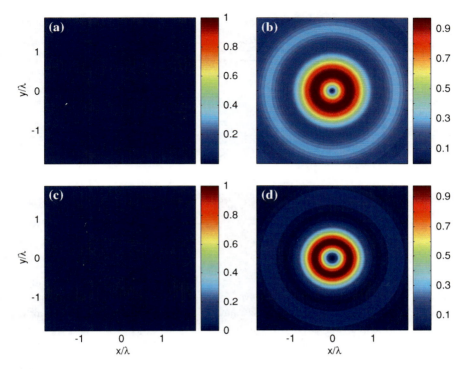

Fig. 4.5 a–c Absolute magnitude of the cylindrical field components (E_ρ, E_φ, E_z) and **d** optical intensity in the focal plane of a lens of NA = 0.95 for azimuthally-polarised illumination. Whilst E_ρ and E_z are identically zero they have been shown for ease of comparison with Fig. 4.4

4.5.2 Focusing of Partially Polarised, Partially Coherent Light

Denoting the CSDM in the focal region of a lens as

$$\mathbb{W}(\boldsymbol{\rho}_1, \boldsymbol{\rho}_2, \omega) \equiv \langle \mathbf{E}(\boldsymbol{\rho}_1, \omega) \mathbf{E}^\dagger(\boldsymbol{\rho}_2, \omega) \rangle, \qquad (4.57)$$

and similarly for the CSDM in other domains in the optical system the problem of focusing spatially inhomogeneous, partially coherent, partially polarised light is now considered (see Table 4.3 for a summary of the notation used for the CSDM and other related quantities at different points in the focusing system).

Given Eq. (4.57) and the scaled Debye-Wolf integral (Eq. (4.45)) it is a simple matter to determine the CSDM for light focused by a lens. Substitution yields

Table 4.3 Summary of notation used when considering focusing of spatially inhomogeneous partially polarised, partially coherent light

	Reference sphere	Back focal plane	Focal region
Coordinates	$\{\theta, \phi\}$	$\{\theta, \phi\}$	$\boldsymbol{\rho} = \{\rho, \varphi, z\}$
Electric field vector	$\mathbf{e}(\theta, \phi)$	$\widetilde{\mathbf{E}}(\theta, \phi)$	$\mathbf{E}(\boldsymbol{\rho})$
Cross-spectral density matrix	$\mathbb{w}(\theta_1, \phi_1, \theta_2, \phi_2)$	$\widetilde{\mathbb{W}}(\theta_1, \phi_1, \theta_2, \phi_2)$	$\mathbb{W}(\boldsymbol{\rho}_1, \boldsymbol{\rho}_2)$
Scalar-based coherent mode	$\psi_n^{(i)}(\theta, \phi)$	$\widetilde{\psi}_n^{(i)}(\theta, \phi)$	$\chi_n^{(i)}(\boldsymbol{\rho})$
Vector-based coherent mode	$\boldsymbol{\Psi}_n(\theta, \phi)$	$\widetilde{\boldsymbol{\Psi}}_n(\theta, \phi)$	$\check{\boldsymbol{\chi}}_n(\boldsymbol{\rho})$

$$\mathbb{W}(\boldsymbol{\rho}_1, \boldsymbol{\rho}_2) = K_1 K_2^* \int_0^{2\pi}\int_0^{2\pi}\int_0^{\alpha}\int_0^{\alpha} \langle \mathbf{e}(\theta_1, \phi_1)\mathbf{e}^{\dagger}(\theta_2, \phi_2)\rangle \exp[ik\Delta_{12}] \\ \times \exp[ik(Z_1\cos\theta_1 - Z_2\cos\theta_2)]\sin\theta_1 \sin\theta_2 d\theta_1 d\theta_2 d\phi_1 d\phi_2, \tag{4.58}$$

where $K_l = -\frac{ikf^2}{2\pi(f+z_l)}\exp(ik\Phi_0)$ and $\Delta_{12} = R_1\sin\theta_1\cos(\phi_1 - \varphi_1) - R_2\sin\theta_2\cos(\phi_2 - \varphi_2)$. Knowledge of the CSDM in a single transverse plane is sufficient to calculate the CSDM on any transverse plane in the focal region via, for example, the Wolf equations [89]. Henceforth, the simplifying assumption that $Z_1 = Z_2 = Z$ will be made, i.e. attention shall be restricted to a single plane in the focal region. Consequently $K_1 = K_2 = K$ also follows.

Finally defining the CSDM on the Gaussian reference sphere in an analogous way to Eq. (4.57), such that $\mathbb{w}(\theta_1, \phi_1, \theta_2, \phi_2) = \langle \mathbf{e}(\theta_1, \phi_1)\mathbf{e}^{\dagger}(\theta_2, \phi_2)\rangle$ gives

$$\mathbb{W}(\boldsymbol{\rho}_1, \boldsymbol{\rho}_2) = |K|^2 \int_0^{2\pi}\int_0^{2\pi}\int_0^{\alpha}\int_0^{\alpha} \mathbb{w}(\theta_1, \phi_1, \theta_2, \phi_2) \exp[ik\Delta_{12}] \\ \times \exp[ikZ(\cos\theta_1 - \cos\theta_2)]\sin\theta_1 \sin\theta_2 d\theta_1 d\theta_2 d\phi_1 d\phi_2. \tag{4.59}$$

In some applications it may be more useful to define the focused CSDM in terms of the CSDM in the back-focal plane of the lens denoted $\widetilde{\mathbb{W}}(\theta_1, \phi_1, \theta_2, \phi_2) = \langle \widetilde{\mathbf{E}}(\theta_1, \phi_1)\widetilde{\mathbf{E}}^{\dagger}(\theta_2, \phi_2)\rangle$. Using this definition and Eq. (4.47) gives

$$\mathbb{w}(\theta_1, \phi_1, \theta_2, \phi_2) = \mathbb{Q}(\theta_1, \phi_1) \cdot \widetilde{\mathbb{W}}(\theta_1, \phi_1, \theta_2, \phi_2) \cdot \mathbb{Q}^{\dagger}(\theta_2, \phi_2). \tag{4.60}$$

Substituting Eq. (4.60) in Eq. (4.59) yields

$$\mathbb{W}(\boldsymbol{\rho}_1, \boldsymbol{\rho}_2) = |K|^2 \int_0^{2\pi}\int_0^{2\pi}\int_0^{\alpha}\int_0^{\alpha} \mathbb{Q}(\theta_1, \phi_1) \cdot \widetilde{\mathbb{W}}(\theta_1, \phi_1, \theta_2, \phi_2) \cdot \mathbb{Q}^{\dagger}(\theta_2, \phi_2) \\ \times \exp[ik\Delta_{12}]\exp[ikZ(\cos\theta_1 - \cos\theta_2)]\sin\theta_1 \\ \times \sin\theta_2 d\theta_1 d\theta_2 d\phi_1 d\phi_2. \tag{4.61}$$

4.5 Focusing of Vectorial Beams

Equations (4.59) and (4.61) are the key results of this section. In what follows various cases are considered, under which these integrals simplify from the four-fold integrals given to separable two-fold integrals (Sect. 4.5.2.1), or even under certain symmetry assumptions, single integrals (Sect. 4.5.2.2).

4.5.2.1 Coherent Mode Representations

Scalar coherent mode expansions in optical coherence theory (see e.g. [89] for a fuller discussion) were perhaps first pioneered by Wolf [86], but have seen fervent use by other authors, e.g. [13, 58, 78]. It should however be noted that all such theories derive from Karhunen-Loéve theory [41, 48] which has been employed in statistics since the 1940s. Karhunen-Loéve theory states that given a (Hermitian, non-negative definite, square integrable) scalar correlation function over a closed domain $\mathcal{D}$, such as the cross-spectral density function $W(\rho_1, \rho_2, \omega)$, it is possible to expand it in terms of an infinite, orthonormal set of coherent modes, $\psi_n(\rho, \omega)$, according to

$$W(\rho_1, \rho_2) = \sum_{n=0}^{\infty} \lambda_n \psi_n^*(\rho_1) \psi_n(\rho_2), \tag{4.62}$$

where the coherent modes and associated expansion coefficients $\lambda_n(\omega)$ are found by solution of the Fredholm integral equation

$$\int_{\mathcal{D}} W(\rho_1, \rho_2) \psi_n(\rho_1) d\rho_1 = \lambda_n \psi_n(\rho_2). \tag{4.63}$$

Extension of existing scalar results to a treatment of the full electromagnetic problem is however more controversial, with two opposing schools of thought debating the appropriate form of coherent mode expansions for 2D (3D) fields. Both of the alternative formalisms, which shall be termed the *scalar-* and *vector based* formalisms respectively, are considered here.

The first, scalar-based interpretation applies the scalar formulation described above to each field component individually, hence requiring the solution of two (three) *uncoupled* Fredholm integral equations of the form of Eq. (4.63). Accordingly the individual elements of a general CSDM are expressed in the form [89]

$$W_{ij}(\rho_1, \rho_2) = \begin{cases} \sum_{n=0}^{\infty} \lambda_n^{(i)} \psi_n^{(i)*}(\rho_1) \psi_n^{(i)}(\rho_2) & \text{for } i = j, \\ \sum_{n=0}^{\infty} \sum_{m=0}^{\infty} \Lambda_{nm}^{(ij)} \psi_n^{(i)*}(\rho_1) \psi_m^{(j)}(\rho_2) & \text{for } i \neq j, \end{cases} \tag{4.64}$$

where $W_{ij}(\rho_1, \rho_2)$ is the (i, j)th element of $\mathbb{W}(\rho_1, \rho_2)$ and the expansion coefficients for off-diagonal terms, $\Lambda_{nm}^{(ij)}$, are found according to the integral

$$\Lambda_{nm}^{(ij)} = \int_D \int_D \psi_n^{(i)}(\rho_1) W_{ij}(\rho_1, \rho_2) \psi_m^{(j)*}(\rho_2) d\rho_1 d\rho_2. \qquad (4.65)$$

Alternatively the vector-based formalism solves the Fredholm integral equation with matrix-valued kernel

$$\int_D \mathbb{W}(\rho_1, \rho_2) \mathbf{\Psi}_n(\rho_1) d\rho_1 = \lambda_n \mathbf{\Psi}_n(\rho_2) \qquad (4.66)$$

to find vectorial coherent modes [23, 70], such that

$$\mathbb{W}(\rho_1, \rho_2) = \sum_{n=0}^{\infty} \lambda_n \mathbf{\Psi}_n(\rho_1) \mathbf{\Psi}_n^{\dagger}(\rho_2). \qquad (4.67)$$

Although this approach is more mathematically involved, since it requires the solution of two (three) *coupled* scalar Fredholm integral equations, it does express the off-diagonal elements more concisely.

Motivated by the analytic advantages frequently afforded by use of coherent mode expansions, they are now used to describe focusing of partially polarised, partially coherent light. Furthermore, expansions of the CSDMs $w(\theta_1, \phi_1, \theta_2, \phi_2)$ on the reference sphere and $\widetilde{\mathbb{W}}(\theta_1, \phi_1, \theta_2, \phi_2)$ in the back focal plane will be considered.

Consider first the scalar-based expansion of $w(\theta_1, \phi_1, \theta_2, \phi_2)$. Using Eq. (4.59) immediately gives

$$W_{ij}(\rho_1, \rho_2) = |K|^2 \sum_{n=0}^{\infty} \lambda_n^{(i)} \int_0^{2\pi} \int_0^{2\pi} \int_0^{\alpha} \int_0^{\alpha} \psi_n^{(i)*}(\theta_1, \phi_1) \psi_n^{(i)}(\theta_2, \phi_2) \exp[ik\Delta_{12}]$$
$$\times \exp[ikZ(\cos\theta_1 - \cos\theta_2)] \sin\theta_1 \sin\theta_2 d\theta_1 d\theta_2 d\phi_1 d\phi_2,$$

for $i = j$ and

$$W_{ij}(\rho_1, \rho_2) = |K|^2 \sum_{n=0}^{\infty} \sum_{m=0}^{\infty} \Lambda_{nm}^{(ij)} \int_0^{2\pi} \int_0^{2\pi} \int_0^{\alpha} \int_0^{\alpha} \psi_n^{(i)*}(\theta_1, \phi_1) \psi_m^{(j)}(\theta_2, \phi_2)$$
$$\times \exp[ik\Delta_{12}] \times \exp[ikZ(\cos\theta_1 - \cos\theta_2)]$$
$$\times \sin\theta_1 \sin\theta_2 d\theta_1 d\theta_2 d\phi_1 d\phi_2,$$

for $i \neq j$. Letting

$$\chi_n^{(i)}(\rho_l) = K \int_0^{2\pi} \int_0^{\alpha} \psi_n^{(i)*}(\theta_l, \phi_l)$$
$$\times \exp[ikR_l \sin\theta_l \cos(\phi_l - \varphi_l)] e^{ikZ\cos\theta_l} \sin\theta_l d\theta_l d\phi_l, \qquad (4.68)$$

4.5 Focusing of Vectorial Beams

gives

$$W_{ij}(\boldsymbol{\rho}_1, \boldsymbol{\rho}_2) = \begin{cases} \sum_{n=0}^{\infty} \lambda_n^{(i)} \chi_n^{(i)*}(\boldsymbol{\rho}_1) \chi_n^{(i)}(\boldsymbol{\rho}_2) & \text{for } i = j, \\ \sum_{n=0}^{\infty} \sum_{m=0}^{\infty} \Lambda_{nm}^{(ij)} \chi_n^{(i)*}(\boldsymbol{\rho}_1) \chi_m^{(j)}(\boldsymbol{\rho}_2) & \text{for } i \neq j. \end{cases} \quad (4.69)$$

For a vector-based expansion of $\mathbb{w}(\theta_1, \phi_1, \theta_2, \phi_2)$ in terms of the set of coherent modes $\boldsymbol{\Psi}_n(\theta, \phi)$ similar results follow, specifically

$$\mathbb{W}(\boldsymbol{\rho}_1, \boldsymbol{\rho}_2) = \sum_{n=0}^{\infty} \lambda_n \, \boldsymbol{\chi}_n(\boldsymbol{\rho}_1) \boldsymbol{\chi}_n^{\dagger}(\boldsymbol{\rho}_2), \quad (4.70)$$

where

$$\boldsymbol{\chi}_n(\boldsymbol{\rho}_l) = K \int_0^{2\pi} \int_0^{\alpha} \boldsymbol{\Psi}_n(\theta_l, \phi_l) \exp\left[i k R_l \sin\theta_l \cos(\phi_l - \varphi_l)\right] e^{ikZ\cos\theta_l} \sin\theta_l d\theta_l d\phi_l. \quad (4.71)$$

Comparing Eqs. (4.69) and (4.70) to the definition of the scalar- and vector-based expansions given by Eqs. (4.64) and (4.67) respectively, it is apparent that the scalar (vector) coherent modes in the focal region can be found by focusing the coherent modes on the reference sphere by use of the scaled Debye-Wolf integral with scalar (vector) kernel. This result is expected because by construction the modes are fully spatially and temporally coherent in addition to being statistically uncorrelated [89]. Consequently each coherent mode can be propagated independently using more familiar ideas from coherent optical theories. It should however be noted that

$$\int_0^{2\pi} \int_0^{\infty} \boldsymbol{\chi}_n^{\dagger}(\boldsymbol{\rho}_l) \boldsymbol{\chi}_m(\boldsymbol{\rho}_l) R_l d R_l d\varphi_l = K_l K_l^* \delta_{nm}, \quad (4.72)$$

meaning that to maintain orthonormality it is necessary to normalise by the factor $|K|$, which yields the alternative, albeit equivalent, expansion

$$\mathbb{W}(\boldsymbol{\rho}_1, \boldsymbol{\rho}_2) = \sum_{n=0}^{\infty} \lambda_n |K|^2 \, \check{\boldsymbol{\chi}}_n(\boldsymbol{\rho}_1) \check{\boldsymbol{\chi}}_n^{\dagger}(\boldsymbol{\rho}_2), \quad (4.73)$$

where $\check{\boldsymbol{\chi}}_n(\boldsymbol{\rho})$ denotes a renormalised coherent mode.

Finally consider coherent mode expansions of the CSDM $\widetilde{\mathbb{W}}(\theta_1, \phi_1, \theta_2, \phi_2)$ in the back focal plane. The scalar and vector based coherent modes are denoted $\widetilde{\psi}_n^{(i)}(\theta, \phi)$ and $\widetilde{\boldsymbol{\Psi}}_n(\theta, \phi)$ respectively. To formulate this problem the coherent modes on the reference sphere need only be related to those in the back focal plane. For the vector-based expansion Eq. (4.60) gives $\boldsymbol{\Psi}_n(\theta, \phi) = \mathbb{Q}(\theta, \phi) \cdot \widetilde{\boldsymbol{\Psi}}_n(\theta, \phi)$. When considering the scalar-based expansion however the mixing of the elements of the CSDM caused by the transformation of Eq. (4.60) means that the focused CSDM cannot be expressed in the form of Eq. (4.69). The lack of a simple, analytic correspondence between the coherent modes in the back focal plane and those in the focal region hence suggests that *a scalar-based coherent mode expansion is unsuitable for focusing in electromagnetic problems*. Consequently only vector-based expansions will be considered in the subsequent derivations.

At this juncture it is convenient to define a number of different metrics, commonly used to describe partially coherent light. There is again much dispute regarding the appropriateness and meaning of these quantities, however such discussions are disregarded here. Instead the implications of focusing in terms of each metric are examined. In particular consider the degree of spectral coherence defined in [6] as

$$|\eta(\rho_1, \rho_2)|^2 = \frac{\text{tr}[\mathbb{W}(\rho_1, \rho_2)]^2}{\text{tr}[\mathbb{W}(\rho_1, \rho_1)]\,\text{tr}[\mathbb{W}(\rho_2, \rho_2)]}, \qquad (4.74)$$

and the degree of spectral coherence defined in [70] as

$$\zeta^2(\rho_1, \rho_2) = \frac{\|\mathbb{W}(\rho_1, \rho_2)\|_F^2}{\text{tr}[\mathbb{W}(\rho_1, \rho_1)]\,\text{tr}[\mathbb{W}(\rho_2, \rho_2)]}, \qquad (4.75)$$

where $\text{tr}[\cdots]$ and $\|\cdots\|_F$ denote the matrix trace and Frobenius norm respectively. Analogous definitions hold for the light before focusing. Since the CSDM will in general change upon focusing, then so too will the associated degrees of spectral coherence η and ζ. Numerical examples of this will be given in Sect. 4.5.3 however it is informative to consider the effective degree of coherence, $\bar{\zeta}$, over the domain $\mathcal{D}$ for a general CSDM, as defined in [70] by

$$\bar{\zeta}^2 = \frac{\int_\mathcal{D} \int_\mathcal{D} \|\mathbb{W}(\rho_1, \rho_2)\|_F^2 d\rho_1 d\rho_2}{\int_\mathcal{D} \text{tr}[\mathbb{W}(\rho_1, \rho_1)]d\rho_1 \int_\mathcal{D} \text{tr}[\mathbb{W}(\rho_2, \rho_2)]d\rho_2}. \qquad (4.76)$$

Before and after focusing $\bar{\zeta}^2$ evaluates to $\sum_{n=0}^\infty \lambda_n^2 / \left[\sum_{n=0}^\infty \lambda_n\right]^2$ and it is therefore possible to conclude that *the effective degree of spectral coherence $\bar{\zeta}$ is unchanged upon focusing*. Unfortunately no conservation rule holds for $\bar{\eta}^2$, which could be defined in an analogous way, again highlighting the question over the appropriateness and meaning of each parameter.

4.5.2.2 Harmonic Angular Dependence

Further simplifications of the focusing integrals of Eqs. (4.59) and (4.61) can be made if certain symmetry conditions hold. In particular the azimuthal integration can be evaluated analytically when the coherent modes (on either the reference sphere or the

4.5 Focusing of Vectorial Beams

back focal plane) have a harmonic angular dependence i.e. $\boldsymbol{\Psi}_n(\theta, \phi) = \boldsymbol{\Psi}_n(\theta) \sin m\phi$ or $\boldsymbol{\Psi}_n(\theta) \cos m\phi$ and similarly for $\widetilde{\boldsymbol{\Psi}}_n(\theta, \phi)$, where m is a non-negative integer.

To consider the assertion of harmonic angular dependence on the reference sphere it is sufficient to consider the $\chi_n(\rho_l)$ integrals of Eq. (4.71) such that

$$\chi_n^I(\rho_l) = K \int_0^{2\pi} \int_0^{\alpha} \boldsymbol{\Psi}_n(\theta_l) \begin{Bmatrix} \sin m\phi_l \\ \cos m\phi_l \end{Bmatrix}$$
$$\times \exp[ikR_l \sin\theta_l \cos(\phi_l - \varphi_l)] e^{ikZ\cos\theta_l} \sin\theta_l d\theta_l d\phi_l. \quad (4.77)$$

Employing Eq. (4.51) gives

$$\chi_n^I(\rho_l) = 2\pi i^m K \begin{Bmatrix} \sin m\varphi_l \\ \cos m\varphi_l \end{Bmatrix} \int_0^{\alpha} \boldsymbol{\Psi}_n(\theta_l) J_m(kR_l \sin\theta_l) e^{ikZ\cos\theta_l} \sin\theta_l d\theta_l. \quad (4.78)$$

Alternatively when considering coherent modes on the back focal plane

$$\chi_n^{II}(\rho_l) = K \int_0^{2\pi} \int_0^{\alpha} \mathbb{Q}(\theta_l, \phi_l) \widetilde{\boldsymbol{\Psi}}_n(\theta_l) \begin{Bmatrix} \sin m\phi_l \\ \cos m\phi_l \end{Bmatrix}$$
$$\times \exp[ikR_l \sin\theta_l \cos(\phi_l - \varphi_l)] e^{ikZ\cos\theta_l} \sin\theta_l d\theta_l d\phi_l. \quad (4.79)$$

Including the explicit form of $\mathbb{Q}(\theta, \phi)$ and analytically performing the integration of ϕ_l gives

$$\chi_n^{II}(\rho_l)$$
$$= \frac{K}{2} \begin{pmatrix} \Theta_{2,-m,2}^{n,x,s} - \Theta_{2,-m,2}^{n,y,c} - i\Theta_{1,-m,3}^{n,z,s} + 2\Theta_{0,m,1}^{n,x,s} + i\Theta_{1,m,3}^{n,z,s} - \Theta_{2,m,2}^{n,x,s} + \Theta_{2,m,2}^{n,y,c} \\ -\Theta_{2,-m,2}^{n,x,c} - \Theta_{2,-m,2}^{n,y,s} + i\Theta_{1,-m,3}^{n,z,c} + 2\Theta_{0,m,1}^{n,y,s} - i\Theta_{1,m,3}^{n,z,c} + \Theta_{2,m,2}^{n,x,c} + \Theta_{2,m,2}^{n,y,s} \\ i\Theta_{1,-m,3}^{n,x,s} - i\Theta_{1,-m,3}^{n,y,c} + 2\Theta_{0,m,4}^{n,z,s} - i\Theta_{1,m,3}^{n,x,s} + i\Theta_{1,m,3}^{n,y,c} \end{pmatrix}$$
(4.80)

for a sinusoidal angular dependence, or

$$\chi_n^{II}(\rho_l)$$
$$= \frac{K}{2} \begin{pmatrix} -\Theta_{2,-m,2}^{n,x,c} - \Theta_{2,-m,2}^{n,y,s} + i\Theta_{1,-m,3}^{n,z,c} + 2\Theta_{0,m,1}^{n,x,c} + i\Theta_{1,m,3}^{n,z,c} - \Theta_{2,m,2}^{n,x,c} - \Theta_{2,m,2}^{n,y,s} \\ -\Theta_{2,-m,2}^{n,x,s} + \Theta_{2,-m,2}^{n,y,c} + i\Theta_{1,-m,3}^{n,z,s} + 2\Theta_{0,m,1}^{n,y,c} + i\Theta_{1,m,3}^{n,z,s} - \Theta_{2,m,2}^{n,xs} + \Theta_{2,m,2}^{n,y,c} \\ -i\Theta_{1,-m,3}^{n,x,c} - i\Theta_{1,-m,3}^{n,y,s} + 2\Theta_{0,m,4}^{n,z,c} - i\Theta_{1,m,3}^{n,x,c} - i\Theta_{1,m,3}^{n,y,s} \end{pmatrix}$$
(4.81)

for a cosinusoidal angular dependence, where

$$\Theta_{q,\pm m,u}^{n,\nu,t}(\rho_l) = 2\pi i^{\pm m} \begin{Bmatrix} \sin(q \pm m)\varphi_l \\ \cos(q \pm m)\varphi_l \end{Bmatrix}$$
$$\times \int_0^\alpha a(\theta_l)\widetilde{\Psi}_n^\nu(\theta_l) p_u \sin\theta_l J_{q\pm m}(kR_l \sin\theta_l) e^{ikZ\cos\theta_l} d\theta_l, \tag{4.82}$$

$\widetilde{\Psi}_n^\nu$ denotes the νth component of $\widetilde{\Psi}_n$ and the sin (cos) term is taken for $t = s$ (c). Evaluation of the single integrals of Eq. (4.82) is all that is necessary to calculate the CSDM of focused, inhomogeneous, partially polarised, partially coherent light for which the coherent modes have a harmonic angular dependence.

In coherence calculations the assumption of a circularly symmetric CSDM is often made (whereby either $w(\theta_1, \phi_1, \theta_2, \phi_2) = w(\theta_1, \theta_2)$ or $\widetilde{W}(\theta_1, \phi_1, \theta_2, \phi_2) = \widetilde{W}(\theta_1, \theta_2)$) because it allows the dimensionality of analysis to be reduced. Circular symmetry in the CSDM is inherited by the coherent modes and hence this frequently considered scenario is given as a special case ($m = 0$) of the preceding analysis. It has however been demonstrated that even under less stringent assumptions the dimensionality of the problem can still be reduced. Finally, it should be noted that in the preceding analysis it was assumed that each field component of the vector based coherent modes had the same harmonic behaviour. This assumption is however not required since Eq. (4.51) can still be used to form a family of integrals similar to that defined by Eq. (4.82). An example of this type is considered in the next section.

4.5.3 Examples

4.5.3.1 Radially Polarised Laguerre-Gauss Modes

By way of example consider a beam-like source formed by the superposition of mutually uncorrelated, radially polarised Laguerre-Gauss modes located in the back focal plane of a lens. Laguerre-Gauss modes, can for example be obtained from laser cavities with circular geometries [9]. In this scenario the CSDM in the back focal plane is of the form $\widetilde{W}(\theta_1, \phi_1, \theta_2, \phi_2) = \sum_{n=0}^\infty \lambda_n \widetilde{\Psi}_n(\theta_1, \phi_1) \widetilde{\Psi}_n^\dagger(\theta_2, \phi_2)$ where

$$\widetilde{\Psi}_n(\theta, \phi) = \widetilde{\Psi}_n(\theta) \begin{pmatrix} \cos\phi \\ \sin\phi \\ 0 \end{pmatrix}, \tag{4.83}$$

and

$$\widetilde{\Psi}_n(\theta) = \left(\frac{2}{\pi\mu^2}\right)^{1/2} L_n\left(\frac{2\sin^2\theta}{\mu^2}\right) \exp\left(-\frac{\sin^2\theta}{\mu^2}\right). \tag{4.84}$$

4.5 Focusing of Vectorial Beams

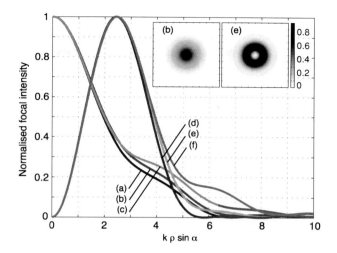

Fig. 4.6 Radial line scans ($\varphi = 0$) and full transverse focused intensity distributions for a radially polarised beam source for **a** $\bar{\zeta} = 0$ ($q = 0$, coherent), **b** $\bar{\zeta} = 1/3$ ($q = 0.62$) and **c** $\bar{\zeta} = 2/3$ ($q = 0.89$). Similar line scans for an azimuthally polarised beam (**d, e** and **f**). For numerical calculations an aplanatic lens of numerical aperture 0.97 was assumed. Furthermore the values $\mu = 1$ and $\lambda = 405$ nm were used. Note that peak intensity has been normalised to unity in all cases for easy comparison

L_n represents the nth order Laguerre polynomial and μ is a frequency dependent parameter. Further consider the case discussed in [24] for which $\lambda_n = \pi(1-q^2)q^{2n}/2\mu^2$ for $0 < q < 1$. The parameter μ is a measure of the beam diameter measured in focal lengths, whilst q determines the effective degree of spectral coherence via $\bar{\zeta}^2 = (1-q^2)/(1+q^2)$, with the limits $q \to 0$ ($q \to 1$) giving a fully spatially (un)correlated source. The beam diameter as specified by μ will be held constant throughout the remainder of this work to avoid extraneous effects resulting from a different apodisation of the beam.

Following the analysis given in Sect. 4.5.2.2 the focused coherent modes are found to be

$$\chi_n^{II}(\rho_l) = K \begin{pmatrix} (\Theta_{0,1,1}^n - \Theta_{2,-1,2}^n) \cos\varphi \\ (\Theta_{0,1,1}^n - \Theta_{2,-1,2}^n) \sin\varphi \\ -i\Theta_{1,-1,3}^n \end{pmatrix}, \tag{4.85}$$

where

$$\Theta_{0,1,1}^n(\rho_l) = 2\pi i \int_0^\alpha a(\theta_l)\widetilde{\Psi}_n(\theta_l)(\cos\theta_l + 1)\sin\theta_l J_1(kR_l\sin\theta_l)e^{ikZ\cos\theta_l}d\theta_l, \tag{4.86}$$

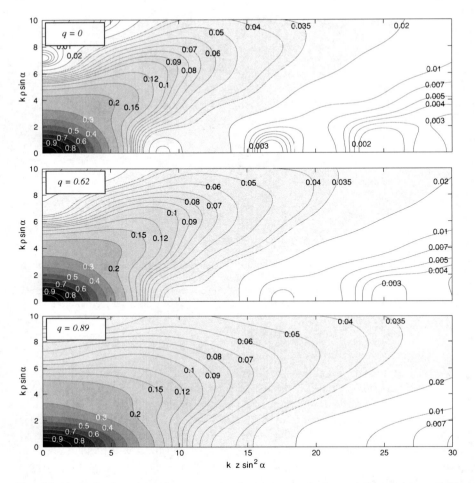

Fig. 4.7 Normalised axial focal intensity distributions ($\varphi = 0$) for partially coherent radially polarised collimated sources with differing effective degrees of spectral coherence as specified by $\bar{\zeta} = 0$ (*top*), $\bar{\zeta} = 1/3$ (*centre*) and $\bar{\zeta} = 2/3$ (*bottom*). Other simulation parameters used were the same as in Fig. 4.6

$$\Theta_{2,-1,2}^{n}(\boldsymbol{\rho}_l) = -2\pi i \int_0^\alpha a(\theta_l)\widetilde{\Psi}_n(\theta_l)(\cos\theta_l - 1)\sin\theta_l J_1(kR_l\sin\theta_l)e^{ikZ\cos\theta_l}d\theta_l, \tag{4.87}$$

$$\Theta_{1,-1,3}^{n}(\boldsymbol{\rho}_l) = -4\pi i \int_0^\alpha a(\theta_l)\widetilde{\Psi}_n(\theta_l)\sin^2\theta_l J_0(kR_l\sin\theta_l)e^{ikZ\cos\theta_l}d\theta_l. \tag{4.88}$$

Using these coherent modes it is possible to calculate the focal intensity distribution for sources of differing effective degree of spectral coherence $\bar{\zeta}$. In Fig. 4.6 transverse line scans ($\varphi = 0$, $Z = 0$) for sources with $\bar{\zeta} = 0$, $1/3$ and $2/3$ are plotted. Although

4.5 Focusing of Vectorial Beams

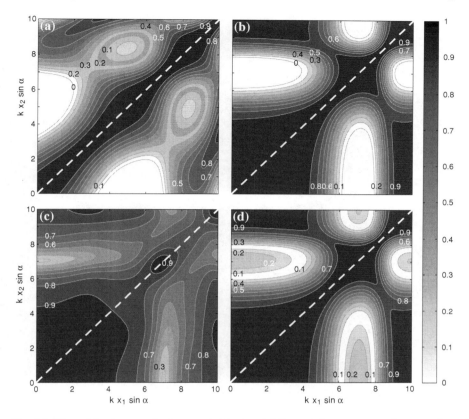

Fig. 4.8 Plots of the degrees of coherence, $|\eta|$ (*top*) and ζ (*bottom*), between two points x_1 and x_2 on the positive x-axis in the focal plane for focused partially coherent radially (*left*) and azimuthally (*right*) polarised collimated light. A value of $q = 0.62$ was used, whilst other simulation parameters were the same as in Fig. 4.6

there is little effect on the width of the transverse profile it is noted with reference to Fig. 4.7 that there is a modest extension in the depth of field as the source becomes more incoherent. There is also a slight increase in the energy density in the wings of the transverse profile. Due to the apodisation over the pupil the focal spot is broader than that for uniform intensity since the contribution from the longitudinal field component, responsible for the narrow spot for the unapodised case, is reduced.

Figure 4.8a, c shows plots of the degrees of spectral coherence, η and ζ, between points located along the positive x-axis ($\phi_1 = \phi_2 = 0$) in the focal plane. Unity degree of coherence between two points implies that were the fields from these points brought together, the resulting interference fringes would have maximum visibility. Consequently if $\rho_1 = \rho_2$ then η automatically evaluates to unity as can be seen along the dashed line in Fig. 4.8a. However this is not in general true for ζ, since this also measures the correlations between individual components of the

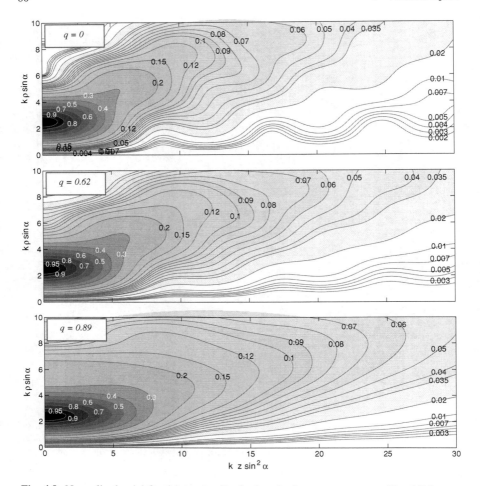

Fig. 4.9 Normalised axial focal intensity distributions in the same manner as Fig. 4.7 for partially coherent azimuthally polarised collimated sources with differing effective degrees of spectral coherence

electric field. The rotation of the electric field vector by a lens can introduce differing stochastic behaviour in orthogonal field components, hence resulting in the possibility of $\zeta(\rho, \rho) \neq 1$ as can be seen along the diagonal in Fig. 4.8. The differences between the two metrics are more fully discussed in [6].

4.5.3.2 Azimuthally Polarised Laguerre-Gauss Modes

For azimuthally polarised beams the vectorial coherent modes are assumed to be of the form,

4.5 Focusing of Vectorial Beams

$$\widetilde{\boldsymbol{\Psi}}_n(\theta, \phi) = \widetilde{\Psi}_n(\theta) \begin{pmatrix} \sin\phi \\ -\cos\phi \\ 0 \end{pmatrix}, \quad (4.89)$$

which in the focal region yields

$$\chi_n^{II}(\rho_l) = K \begin{pmatrix} (\Theta_{0,1,1}^n + \Theta_{2,-1,2}^n) \sin\varphi \\ -(\Theta_{0,1,1}^n + \Theta_{2,-1,2}^n) \cos\varphi \\ 0 \end{pmatrix}. \quad (4.90)$$

Again transverse line scans of the focal intensity distribution are shown in Fig. 4.6, whilst the axial intensity distribution is shown in Fig. 4.9. Similar conclusions to those made for the radially polarised source can be drawn for an azimuthally polarised source, however the augmentation of the wings of the transverse intensity profile is more pronounced. Figure 4.8b, d again shows plots of the degrees of spectral coherence, η and ζ, between points located along the positive x-axis in the focal plane. For azimuthally polarised illumination the resulting plots are very similar, because along the x-axis the focused coherent modes are purely y-polarised.

4.6 Conclusions

Having started from the foundation of electromagnetism as encapsulated in Maxwell's equations, this chapter set out to detail many facets of vectorial optics. The polarisation state of optical waves has been defined and numerous alternative representations considered, from which it was found each has its own merits and flaws. In particular, representations such as Jones vectors are incapable of fully specifying the nature of statistical sources. In this vein, further representations stemming from a parameterisation of the second order source statistics were introduced, including the coherency and Stokes vectors. With an eye to requirements in focusing systems in which the field in the focal region of a lens is fundamentally 3D in nature the conventional 2D treatments were extended to describe 3D fields.

Similarly, two algebraic frameworks within which polarised light can be propagated through non-depolarising and depolarising optical systems were discussed, including the 3D extension, namely those of Jones and Mueller. The generalised Jones matrix formalism is unfortunately intrinsically a geometric ray method (i.e. diffraction effects are neglected) and hence only approximate in many scenarios. However, it was also seen that this formalism can frequently be used to rigorously calculate boundary conditions for diffraction integrals, which allow the full vectorial wave nature of light to be modelled. As such, new work by the author was detailed in which a general description of focusing of partially polarised, partially coherent electromagnetic waves capable of handling spatially inhomogeneous statistical properties across the pupil of the focusing lens(es) was developed. This was achieved by use of the scaled Debye-Wolf diffraction integral, which places few constraints on the system geometry since it is valid for high NA lenses of arbitrary Fresnel number.

Computationally, the integration routines required to focus beams with arbitrary coherence and polarisation properties are expensive. By employing a coherent mode representation of the CSDM it is however possible to reduce the four-dimensional integrals to two-dimensional ones allowing substantial computational gains to be made. Analysis of the focusing operation was performed in terms of the CSDM across both the Gaussian reference sphere and the back focal plane in terms of scalar- and vector-based coherent modes, since both are frequently used in optical calculations. It was found however that due to mixing of different components of the electric field that occurs in high NA optical systems scalar-based coherent mode expansions can be unsuitable. It was also shown that the effective degree of spectral coherence $\bar{\zeta}$ of an electromagnetic beam is unchanged upon focusing, although no equivalent result can be found for the alternative metric found in the literature, namely $\bar{\eta}$.

The imposition of circular symmetry is often made in the analysis of optical systems to make calculations more mathematically tractable and to reduce the dimensionality of the problem. Via the coherent mode expansions detailed these benefits are still realisable with the less stringent requirement of a harmonic angular dependence, as was highlighted by a number of examples.

As a final comment, although this chapter has concentrated solely on the second-order statistical properties as encapsulated by the CSDM, it is in principle possible to extend Eqs. (4.59) and (4.61) to calculate higher order statistical moments of focused light. Since knowledge of all the moments of a random process provides a full description of the process it is thus possible to fully account for the effect of focusing on randomly fluctuating, electromagnetic waves, within the framework discussed.

References

1. R.M.A. Azzam, N.M. Bashara, *Ellipsometry and Polarised Light* (Elsevier, Amsterdam, 1987)
2. R. Bartholinus, *Experimenta Crystalli Islandici Disdiaclastici Quibus Mira and Insolita Refractio Detegitur* (Danielis Paulli, Hafniæ, 1669)
3. M.L. Boas, *Mathematical Methods in the Physical Sciences*, 2nd edn. (Wiley, New York, 1983)
4. M. Born, E. Wolf, *Principles of Optics*, 7th edn. (Cambridge University Press, Cambridge, 1980)
5. G. Brooker, *Modern Classical Optics* (Oxford University Press, Oxford, 2003)
6. D.P. Brown, A.K. Spilman, T.G. Brown, R. Borghi, S.N. Volkov, E. Wolf, Spatial coherence properties of azimuthally polarized laser modes. Opt. Commun. **281**, 5287–5290 (2008)
7. R.A. Chipman, *Handbook of Optics*, vol. 2 (McGraw Hill, New York, 1995)
8. R.A. Chipman, J.E. Stacy, *PMAP: polarization matrix analysis program*, COSMIC program NPO-17273 (University of Georgia, Athens, 1983)
9. C.C. Davis, *Lasers and Electro-optics—Fundamentals and Engineering* (Cambridge University Press, Cambridge, 1996)
10. R. Dorn, S. Quabis, G. Leuchs, Sharper focus for a radially polarized light beam. Phys. Rev. Lett. **91**, 233901 (2004)

References

11. M. Faraday, On the magnetization of light and the illumination of magnetic lines of force. Philos. Trans. R. Soc. Lond. **1**, 104–123 (1846)
12. D. Fischer, T. Visser, Spatial correlation properties of focused partially coherent light. J. Opt. Soc. Am. A **21**, 2097–2102 (2004)
13. S. Flewett, H. Quiney, C. Tran, K. Nugent, Extracting coherent modes from partially coherent wavefields. Opt. Lett. **34**, 2198–2200 (2009)
14. M.R. Foreman, P. Török, Focusing of spatially inhomogeneous partially coherent, partially polarized electromagnetic fields. J. Opt. Soc. Am. A **26**, 2470–2479 (2009)
15. A.M. Fox, *Optical Properties of Solids* (Oxford University Press, Oxford, 2002)
16. A. Friberg, Partial polarisation and coherence in arbitrary electromagnetic fields. AIP Conf. Proc. **992**, 460 (2008)
17. D. Ganic, X. Gan, M. Gu, Focusing of doughnut laser beams by a high numerical-aperture objective in free space. Opt. Express **11**, 2747–2752 (2003)
18. W. Gao, Effects of coherence and vector properties of the light on the resolution limit in stimulated emission depletion fluorescence microscopy. J. Opt. Soc. Am. A **25**, 1378–1382 (2008)
19. J.J. Gil, Polarimetric characterization of light and media. Eur. Phys. J. Appl. Phys. **40**, 1–47 (2007)
20. D. Goldstein, *Polarised Light: Fundamentals and Applications* (Marcel Dekker, New York, 2003)
21. J.W. Goodman, *Introduction to Fourier Optics*, 2nd edn. (McGraw Hill, New York, 1996)
22. J.W. Goodman, *Statistical Optics* (Wiley, New York, 2004)
23. F. Gori, M. Santarsiero, R. Simon, G. Piquero, R. Borghi, G. Guattari, Coherent-mode decomposition of partially polarized, partially coherent sources. J. Opt. Soc. Am. A **20**, 78–84 (2003)
24. G. Guattari, C. Palma, C. Padovani, Cross-spectral densities with axial symmetry. Opt. Commun. **73**, 173–178 (1989)
25. A. Hardy, D. Treves, Structure of the electromagnetic field near the focus of a stigmatic lens. J. Opt. Soc. Am. **63**, 85–90 (1973)
26. S.W. Hell, J. Wichmann, Breaking the diffraction resolution limit by stimulated emission: stimulated-emission-depletion fluorescence microscopy. Opt. Lett. **19**, 780–782 (1994)
27. P.D. Higdon, P. Török, T. Wilson, Imaging properties of high aperture multiphoton fluorescence scanning optical microscopes. J. Microsc. **193**, 127–141 (1998)
28. C. Huygens, *Traité de la Lumière* (Pieter van der Aa, Leyden, 1690)
29. V.S. Ignatowsky, Diffraction by a lens of arbitrary aperture. Trans. Opt. Inst. Pet. **1**, 1–36 (1919)
30. D.J. Innes, A.L. Bloom, Design of optical systems for use with laser beams. Spectr. Phys. Laser Tech. Bull. **5**, 1–10 (1966)
31. S. Inoué, Studies of depolarisation of light at microscope lens surfaces. II: The origin of stray light by rotation at the lens surfaces. Exp. Cell Res. **3**, 199–208 (1952)
32. J.D. Jackson, *Classical Electrodynamics* (Wiley, New York, 1998)
33. R.C. Jones, A new calculus for the treatment of optical systems: I: Description and discussion of the calculus. J. Opt. Soc. Am. **31**, 488–493 (1941)
34. R.C. Jones, A new calculus for the treatment of optical systems: II: Proof of three general equivalence theorems. J. Opt. Soc. Am. **31**, 493–499 (1941)
35. R.C. Jones, A new calculus for the treatment of optical systems: III: The Sohncke theory of optical activity. J. Opt. Soc. Am. **31**, 500–503 (1941)
36. R.C. Jones, A new calculus for the treatment of optical systems: IV. J. Opt. Soc. Am. **32**, 486–493 (1942)
37. R.C. Jones, A new calculus for the treatment of optical systems: V: A more general formulation and description of another calculus. J. Opt. Soc. Am. **37**, 107–110 (1947)
38. R.C. Jones, A new calculus for the treatment of optical systems: VI: Experimental determination of the matrix. J. Opt. Soc. Am. **37**, 110–112 (1947)
39. R.C. Jones, A new calculus for the treatment of optical systems: VII: Properties of the n-matrices. J. Opt. Soc. Am. **38**, 110–112 (1948)

40. R.C. Jones, A new calculus for the treatment of optical systems: VIII: Electromagnetic theory. J. Opt. Soc. Am. **46**, 126–131 (1956)
41. K. Karhunen, Über lineare methoden in der wahrscheinlichkeitsrechnung. Ann. Acad. Sci. Fennicae Ser. A. I. Math.-Phys. A **137**, 1–79 (1947)
42. Y. Kato, K. Mima, N. Miyanaga, S. Arinaga, Y. Kitagawa, M. Nakatsuka, C. Yamanaka, Random phasing of high-power lasers for uniform target acceleration and plasma-instability suppression. Phys. Rev. Lett. **53**, 1057–1060 (1984)
43. K. Kim, L. Mandel, E. Wolf, Relationship between Jones and Mueller matrices for random media. J. Opt. Soc. Am. A **4**, 433–437 (1987)
44. H. Kubota, S. Inoué, Diffraction images in the polarisation microscope. J. Opt. Soc. Am. **49**, 191–198 (1959)
45. R.H. Lehmberg, A.J. Schmitt, S.E. Bodner, Theory of induced spatial incoherence. J. Appl. Phys. **62**, 2680–2701 (1987)
46. K. Lindfors, T. Setälä, M. Kaivola, A. Friberg, Degree of polarization in tightly focused optical fields. J. Opt. Soc. Am. A **22**, 561–568 (2005)
47. X. Liu, J. Pu, Focal shift and focal switch of partially coherent light in dual-focus systems. Opt. Commun. **252**, 262–267 (2005)
48. M. Loève, *Probability Theory*, 4th edn. (Springer, Heidelberg, 1978)
49. A. Luis, Degree of polarization for three-dimensional fields as a distance between correlation matrices. Opt. Commun. **253**, 10–14 (2005)
50. L. Mandel, E. Wolf, *Optical Coherence and Quantum Optics* (Cambridge University Press, Cambridge, 1995)
51. M. Mansuripar, Distribution of light at and near the focus of high numerical aperture objectives. J. Opt. Soc. Am. A **3**, 2086–2093 (1986)
52. J.C. Maxwell, A dynamical theory of the electromagnetic field. Philos. Trans. R. Soc. Lond. **155**, 459–512 (1865)
53. J.P. McGuire Jr., R.A. Chipman, Diffraction image formation in optical systems with polarization aberrations: I: Formulation and example. J. Opt. Soc. Am. A **7**, 1614–1626 (1990)
54. J.P. McGuire Jr., R.A. Chipman, Diffraction image formation in optical systems with polarization aberrations: II: Amplitude response matrices for rotationally symmetric systems. J. Opt. Soc. Am. A **8**, 833–840 (1991)
55. J.P. McGuire Jr., R.A. Chipman, Polarization aberrations: 1: Rotationally symmetric optical systems. Appl. Opt. **33**, 5080–5100 (1994)
56. H. Mueller, The foundations of optics. J. Opt. Soc. Am. A **8**, 661 (1938)
57. I. Netwon, *Optiks* (Smith and Walford, London, 1704)
58. C. Palma, G. Cincotti, Imaging of j_0 correlated Bessel-Gauss beams. IEEE J. Quantum Electron. **33**, 1032–1040 (1997)
59. N.G. Parke, Optical algebra. J. Math. Phys. **28**, 131 (1949)
60. H. Poincaré, *Théorie Mathématique de la Lumière*, vol. 2, chap. 12 (Gauthiers-Villars, Paris, 1892)
61. S.K. Rhodes, K.A. Nugent, A. Roberts, Precision measurement of the electromagnetic fields in the focal region of a high-numerical-aperture lens using a tapered fiber probe. J. Opt. Soc. Am. A **19**, 1689–1693 (2002)
62. B. Richards, E. Wolf, Electromagnetic diffraction in optical systems: II: Structure of the image field in an aplanatic system. Trans. Opt. Inst. Pet. **253**, 358–379 (1959)
63. C. Rydberg, First- and second-order statistics of partially coherent, high-numerical-aperture optical fields. Opt. Lett. **33**, 104–106 (2008)
64. T. Setälä, J. Tervo, A. Friberg, Complete electromagnetic coherence in the space-frequency domain. Opt. Lett. **29**, 328–330 (2004)
65. W.A. Shurcliff, *Polarised Light: Production and Use* (Harvard University Press, Cambridge, 1962)
66. J.J. Stamnes, *Waves in Focal Regions* (Adam Hilger, Bristol, 1986)
67. G.G. Stokes, On the composition and resolution of streams of polarized light from different sources. Trans. Camb. Phil. Soc. **9**, 399 (1852)

References

68. J. Stolze, D. Suter, *Quantum Computing : A Short Course from Theory to Experiment* (Wiley GmbH, Weinheim, 2004)
69. J.A. Stratton, *Electromagnetic Theory* (McGraw-Hill, New York, 1941)
70. J. Tervo, T. Setälä, A. Friberg, Theory of partially coherent electromagnetic fields in the space-frequency domain. J. Opt. Soc. Am. A **21**, 2205 (2004)
71. P. Török, An imaging theory for advanced, high numerical aperture optical microscopes. DSc Thesis, 2003
72. P. Török, P.D. Higdon, T. Wilson, On the general properties of polarised light conventional and confocal microscopes. Opt. Commun. **148**, 300–315 (1998)
73. P. Török, P.R.T. Munro, The use of Gauss-Laguerre vector beams in STED microscopy. Opt. Express **12**, 3605–3617 (2004)
74. P. Török, P. Varga, Electromagnetic diffraction of light focused through a stratified medium. Appl. Opt. **36**, 2305–2312 (1997)
75. P. Török, P. Varga, G.R. Booker, Electromagnetic diffraction of light focused through a planar interface between materials of mismatched refractive indices: structure of the electromagnetic field: I. J. Opt. Soc. Am. A **12**, 2136–2144 (1995)
76. P. Török, P. Varga, Z. Laczik, G.R. Booker, Electromagnetic diffraction of light focused through a planar interface between materials of mismatched refractive indices: an integral representation. J. Opt. Soc. Am. A **12**, 325–332 (1995)
77. J.S. Tyo, Design of optimal polarimeters: maximization of signal-to-noise ratio and minimization of systematic error. Appl. Opt. **41**, 619–630 (2002)
78. T. van Dijk, G. Gbur, T. Visser, Shaping the focal intensity distribution using spatial coherence. J. Opt. Soc. Am. A **22**, 575–581 (2008)
79. E. Waluschka, Polarization ray trace. Opt. Eng. **28**, 86–89 (1989)
80. W. Wang, A. Friberg, E. Wolf, Focusing of partially coherent light in systems of large Fresnel numbers. J. Opt. Soc. Am. A **14**, 491–496 (1997)
81. G.N. Watson, *A Treatise on the Theory of Bessel Functions* (Cambridge University Press, Cambridge, 1995)
82. N. Wiener, Coherency matrices and quantum theory. J. Math. and Phys. **7**, 109 (1928)
83. N. Wiener, Harmonic analysis and the quantum theory. J. Franklin Inst. **207**, 525 (1929)
84. E. Wolf, Coherence properties of partially polarized electromagnetic radiation. Nuovo Cimento **15**, 1165 (1959)
85. E. Wolf, Electromagnetic diffraction in optical systems: I: An integral representation of the image field. Proc. R. Soc. Lond. A **253**, 349–357 (1959)
86. E. Wolf, New theory of partial coherence in the space-frequency domain: I: Spectra and cross spectra of steady-state sources. J. Opt. Soc. Am. **72**, 343–351 (1982)
87. E. Wolf, *Selected Works of Emil Wolf with Commentary* (World Scientific, Singapore, 2001)
88. E. Wolf, Unified theory of coherence and polarization of random electromagnetic beams. Phys. Lett. A **312**, 263–267 (2003)
89. E. Wolf, *Introduction to the Theory of Coherence and Polarization of Light* (Cambridge University Press, Cambridge, 2007)
90. Z. Zhang, J. Pu, X. Wang, Focusing of partially coherent Bessel-Gaussian beams through a high-numerical-aperture objective. Opt. Lett. **33**, 49–51 (2008)

Chapter 5
Information in Polarimetry

> *Measure what is measurable, and make measurable what is not so.*
>
> Galileo Galilei

5.1 Polarimetry

Polarimetry is the study and measurement of the polarisation state of light and is a popular and useful tool in science today. Applications vary from astronomy, microscopy and biomedical diagnosis [59, 65] to more fundamental crystallographic, material and single molecule studies [20, 74]. Polarisation can also be utilised in quantum cryptography and communication [73]. Although measurement of the state of polarisation of light is often an important objective [29, 78] such polarimetric techniques are also frequently used to obtain information about an optical system, such as its birefringence [13]. One may then subdivide polarimetry into two broad categories: Stokes polarimetry and Mueller polarimetry. The former entails measuring the four Stokes parameters of light, whilst the latter is intended to measure the full Mueller matrix of a sample from which parameters of interest can then be inferred.

During the course of this chapter the performance characteristics of both Stokes and Mueller polarimeters will be investigated and optimised. After a brief introduction to the system model of Stokes and Mueller polarimeters, the concept of polarisation resolution will be introduced and formally quantified in Sect. 5.2. Polarisation resolution in turn allows further performance metrics to be defined, such as the degrees of freedom (or accuracy) in a polarimetric measurement, as may be appropriate for characterisation of polarisation multiplexed systems, and the efficiency of observation, which quantifies the light levels required to achieve a given resolution. The efficiency of observation is particularly pertinent when performing polarimetric measurements with a limited photon budget, as can arise for example in astronomy

Fig. 5.1 Block diagram of a polarimetric measurement

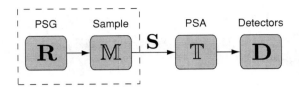

or single molecule studies. A number of existing polarimeter architectures are then assessed in Sect. 5.2.3 using these new informational metrics.

Section 5.3 proceeds to use the proposed figures of merit as a grounds for optimisation of polarimeters. Furthermore, by use of the Bayesian methods introduced in Chap. 3, it is possible to complement the optimisation procedure with probabilistic a priori information. Again a number of examples, including a polarimetric matched filter and a linear polarimeter, are given to illustrate the algorithm in Sect. 5.3.1.

The optimisation framework is detailed in terms of the estimation of the Stokes vector of the incident light or Mueller matrix of a sample. Extension to inference of alternative polarimetric parameters, such as birefringence, is however considered in Sect. 5.3.2, and can also be used to describe the inherent noise propagation and amplification in signal processing. Illustration of the latter is provided by considering the popular Lu–Chipman polar decomposition of a Mueller matrix in Sect. 5.4.

5.1.1 Stokes Polarimetry

When polarised light is transmitted through a polarisation state analyser (PSA) the transmitted intensity D can be found by projecting the input Stokes vector onto a measurement vector $\mathbf{T} = (1, T_1, T_2, T_3)^T/2$ ($\mathbf{T} \in \mathcal{S}$), as defined by the analyser configuration, whereby $D = \mathbf{T}^T \cdot \mathbf{S}$. Here $\mathbf{T}$ is normalised such that $T_1^2 + T_2^2 + T_3^2 = 1$ so that the transmitted intensity $0 \leq D \leq S_0$ i.e. to ensure the analyser is passive. With prior knowledge of the analysing state, as can be deduced from the polarisation elements present in the PSA, it would theoretically be possible to estimate the Stokes vector of the incident light. In general however, multiple measurements are made to improve accuracy and remove ambiguities that may exist. For example a division of amplitude polarimeter (DOAP), as originally proposed by Azzam [2], uses at least four different analysers simultaneously, whilst a null ellipsometer uses an analyser which is varied between sequential measurements [3]. Arranging the N_D respective transmitted intensities into an intensity vector $\mathbf{D} = (D_1, D_2, \ldots, D_{N_D})$, the series of measurements can be described by the matrix equation

$$\mathbf{D} = \mathbb{T}\mathbf{S}, \tag{5.1}$$

where $\mathbb{T}$ is a $N_D \times 4$ instrument matrix with rows corresponding to the N_D measurement states (see Fig. 5.1).

5.1 Polarimetry

To ensure that Eq. (5.1) can describe both sequential or simultaneous measurements it is necessary to introduce a further diagonal matrix $\mathbb{V}$ to account for the beam splitting required for simultaneous measurements such that $\mathbf{D} = \mathbb{V}\mathbb{T}\mathbf{S}$. Conservation of energy (assuming ideal optical elements) dictates that for multiple simultaneous measurements $\text{tr}[\mathbb{V}] = 1$. Alternatively if $\mathbf{D}$ is formed from N_D sequential measurements then $\mathbb{V} = \mathbb{I}$ where $\mathbb{I}$ is the $N_D \times N_D$ identity matrix. Realistically, energy will be lost during propagation through an optical system from absorption and scattering for example. Consequently $\text{tr}[\mathbb{V}]$ can then be used as a measure of the light efficiency of the PSA.

Given a set of noiseless intensities it is possible to deduce the state of polarisation of the incident light by application of the inverse operation i.e.

$$\mathbf{S} = [\mathbb{V}\mathbb{T}]^+ \mathbf{D}, \tag{5.2}$$

where + denotes the Moore-Penrose pseudoinverse [51, 61] of a matrix. In the presence of noise the pseudoinverse given by Eq. (5.2) gives the minimum least square error estimator. For Gaussian noise the minimum least square error estimator is equivalent to the the MLE [38], however this correspondence is not maintained for Poisson noise. The MLE must then be calculated as discussed in Sect. 3.2.2.2, although use of Eq. (5.2) still gives estimates of $\mathbf{S}$ with reasonable accuracy.

Although division of wavefront polarimeters (DOWPs) [14] can also be used for polarimetric measurements this arrangement is neglected in this work since it requires beams that are uniformly polarised and that the beam intensity profile be known a priori; conditions which are generally not achieved in practice. DOWPs are hence rarely used.

5.1.2 Mueller Polarimetry

A Mueller polarimeter builds on the principle of a Stokes polarimeter by addition of a polarisation state generator (PSG) and a sample to the optical setup as shown in Fig. 5.1. The action of the sample on the incident polarised light can be described by a 4×4 Mueller matrix $\mathbb{M}$ such that the light incident into the DOAP is described by the Stokes vector

$$\mathbf{S} = \mathbb{M}\mathbf{R}, \tag{5.3}$$

where $\mathbf{R}$ is the Stokes vector of the illuminating beam. Since the Mueller matrix has 16 elements, all of which must in general be determined, it is not sufficient to illuminate using a single polarisation state. At minimum four distinct polarisation states must be used to illuminate the sample, so as to give the necessary set of 16 simultaneous equations. In general $\mathbf{R}$ can thus be written as a $4 \times N_R$ matrix, $\mathbb{R}$, $N_R \geq 4$ whose columns are the Stokes vectors of the input states. Consequently $\mathbf{D}$ becomes a $N_D \times N_R$ matrix, $\mathbb{D}$, whose columns correspond to the vector of detected intensities for each input polarisation state. Hence

$$\mathbb{D} = \mathbb{V}\mathbb{T}\mathbb{M}\mathbb{R}, \tag{5.4}$$

from which the Mueller matrix can be found using the inverse operations, i.e.

$$\mathbb{M} = [\mathbb{V}\mathbb{T}]^+ \mathbb{D} \mathbb{R}^+. \tag{5.5}$$

5.2 Polarisation Resolution

Spatial resolution in optical imaging systems is a well known and frequently employed concept in system design and evaluation. Multiple metrics of spatial resolution have been defined ranging from the more traditional Abbe (Rayleigh) criterion for (in)coherent optical systems,[1] or the highest transmitted spatial frequency [30], to more recent, specialised metrics, such as localisation accuracy which quantifies the smallest estimable distance between two point sources [58, 64]. A resolution metric for polarimetric systems is however lacking in the literature. Accordingly this section employs the ideas introduced in Chap. 3 to formulate a suitable definition.

Resolution, by any definition, aims to encapsulate the limits to which something can be measured. Consider for example Rayleigh's resolution criterion, which quantifies the minimum angular separation between which two point objects, imaged by an incoherent diffraction limited imaging system, appear separated. Specifically, Rayleigh's criterion states that this minimum separation occurs when the first minimum in the image of one point (an Airy pattern for a circularly symmetric system with unobstructed pupil) coincides with the maximum of the image of the second point object. Although theoretically in a noiseless system two point objects can be infinitely resolved, for example by deconvolving the image with the point spread function of the system, Rayleigh's criterion implicitly (although arbitrarily) makes an assumption of the minimum change of intensity that can be measured, and hence the extent to which the central depression in the total intensity profile can be distinguished. Accordingly, separations smaller than that given by Rayleigh's criterion are said to be unmeasurable. The resolution of a system is therefore synonymous with a region of uncertainty in measurement space.

In Chap. 3 ellipsoids of concentration were introduced in statistical inference problems. The enclosed hyper-volume of these ellipsoids can be used to define a resolution in the Hilbert space associated with the estimation process. Specialising to polarimetry, the measurement problem can be considered as the estimation of a position in the 4D Stokes (or an analogous 16D Mueller) space. To introduce polarisation resolution it is sufficient to consider only Stokes polarimetry, since the mathematical formulation for Mueller polarimetry is identical (although higher dimensional in nature) and hence provides no further insight. Using Eq. (3.31), with $N_w = 4$ and $V_4 = \pi^2/2$, i.e. the volume of the 4D unit hyper-sphere, the resolution in Stokes space is given by

---

[1] Abbe's criterion is also applicable to incoherent systems.

5.2 Polarisation Resolution

$$V_{\mathbf{w}} = \frac{\pi^2}{2}\sqrt{\frac{c^4}{|\mathbb{J}_{\mathbf{w}}^r|}}, \tag{5.6}$$

where $\mathbb{J}_{\mathbf{w}}^r$ is the FIM associated with estimation of the parameter vector $\mathbf{w} = (\check{S}_1, \check{S}_2, \check{S}_3, S_0)$ (the reordering is for later convenience). A suitable choice of c in this definition is no less arbitrary than, for example, the Rayleigh criterion, as it is merely a measure of what is acceptable to the end user. A value of 0.9 will henceforth be assumed.

Typically $\mathbb{J}_{\mathbf{w}}^r$ will be dependent on the state of polarisation being measured, however, when performing a measurement the state of polarisation is generally not known in advance. As a result $\mathbf{w}$ must be treated as a vector random variable and a Bayesian approach adopted. The uncertainty of the experimenter can be modelled by means of a non-informative prior PDF, i.e. by assuming that each state of polarisation of light is equally likely.[2] A Bayesian polarisation resolution $\widetilde{V}_{\mathbf{w}}$ is then defined analogously to Eq. (5.6) in which the BFIM $\mathbb{J}_{\mathbf{w}} = E_{\mathbf{w}}[\mathbb{J}_{\mathbf{w}}^r]$ is used. The tilde notation will be used here to represent metrics derived from the BFIM.

In polarimetry however the absolute intensity of the light may be of secondary or little importance, since the state of polarisation is fully specified by the three variables $\{\check{s}_1, \check{s}_2, \check{s}_3\}$.[3] The estimation problem thus reduces to inferring a position in the 3-dimensional Hilbert space spanned by the parameter vector $\mathbf{u} = (\check{s}_1, \check{s}_2, \check{s}_3)$ corresponding to Poincaré space. Unfortunately, this process still necessitates estimation of the total intensity of the beam s_0, be it implicit or explicit and thus s_0 is considered to be a nuisance parameter.

Ultimately the requisite estimation of s_0 reduces the polarisation accuracy obtainable in Poincaré space, the extent of which can be assessed by partitioning the FIM in an analogous way to that discussed in Sect. 3.3.1 viz.

$$\mathbb{J}_{\mathbf{w}}^r = \begin{pmatrix} \mathbb{J}_{11} & \mathbf{J}_{12} \\ \mathbf{J}_{12}^T & J_{22} \end{pmatrix}, \tag{5.7}$$

where $\mathbb{J}_{11}$ is a 3×3 reduced FIM, $\mathbf{J}_{12}$ is a 3×1 column vector describing the cross-correlations between estimates of $\mathbf{u}$ and s_0, and J_{22} is a scalar, whose reciprocal describes the accuracy achievable for any estimate of s_0 via the CRLB. For any single state of polarisation $V_{\mathbf{w}}/V_{\mathbf{u}} = (3\pi/8)\sqrt{c/|J_{22}|}$ as follows from Eq. (3.32).

The BFIM relevant to estimation of $\mathbf{u}$ is then given by

$$\mathbb{J}_{\mathbf{u}} = E_{\mathbf{u}|s_0}\left[\mathbb{J}_{\mathbf{u}}^r\right] = E_{\mathbf{u}|s_0}\left[\mathbb{J}_{11} - \mathbf{J}_{12}^T J_{22}^{-1} \mathbf{J}_{12}\right], \tag{5.8}$$

where $E_{\mathbf{u}|s_0}[\ldots]$ denotes averaging with respect to $\mathbf{u}$ for a given s_0. The cross-correlations between estimates of $\mathbf{u}$ and s_0 are the cause of the reduction in the

---

[2] The assumption of maximal ignorance will be relaxed in the examples of Sect. 5.3.1.
[3] The convention of denoting a random variable and the value of a particular realisation by an upper and lower case letter as introduced in Chap. 2 has again been used.

polarisation resolution. The reduced Bayesian polarisation resolution in Poincaré space is thus given by

$$\widetilde{V}_{\mathbf{u}} = \frac{4\pi}{3}\sqrt{\frac{c^3}{|\mathbb{J}_{\mathbf{u}}|}}. \tag{5.9}$$

By partitioning $\mathbb{J}_{\mathbf{w}}^r$ in different ways the treatment can be extended to situations in which not all polarisation parameters are desired (see Sect. 3.3.1). Consideration of the reduced BFIM $\mathbb{J}_{\mathbf{u}}$ is insightful since it separates the dependence of the noise on the input intensity and the input state of polarisation.

5.2.1 Polarisation Encoding and Degrees of Freedom

Multiplexing of an optical signal, whereby information is encoded using different degrees of freedom of light, provides a means to increase information storage and transmission rates. For example different wavelengths can be used to send multiple signals along optical fibres [39] in so-called wavelength-division multiplexing (WDM). Fundamentally for WDM the number of different wavelengths, (or more generally the number of channels or degrees of freedom) depends on the bandwidth of the channel and the extent of interchannel interference (crosstalk) that can be tolerated. For example, in fibre optic telecommunication networks which operate in the 1480–1600 nm low loss window of silica glass, the international recommendation is for a wavelength spacing of 0.8 nm ranging from 1537 to 1563 nm, so as to give 32 channels with acceptable levels of crosstalk [37].

Degrees of freedom of information channels were perhaps first considered in the analysis of telegraph signals by Nyquist [56, 57], Küpfmüller [41] and Hartley [34]. Their respective formulations essentially characterised a signal with an (approximately) finite bandwidth and temporal extent using the product of the bandwidth and the duration to quantify the number of degrees of freedom. This was put on a more formal basis by Gabor using Fourier analysis in which the domain of the signal in the frequency-time plane was partitioned into "information cells" of unit area [28]. Similar ideas were later employed in an imaging context in which case the space-bandwidth product was defined and used (see Sect. B.1.1 for further details) [22, 44, 46, 75]. Other definitions also employ the norm of the associated system operator [26, 50, 55, 62].

Polarisation encoding is also possible, however is almost exclusively considered in the context of only two orthogonal states of polarisation [17, 21, 31, 40]. Such analysis is perhaps natural in the sense that crosstalk between the two degrees of freedom is zero in the ideal case, however it automatically forsakes the possibilities afforded by encoding over the entirety of Poincaré space (or Stokes space if amplitude modulation is also employed). Given the ability of polarimeters to distinguish multiple states of polarisation it is hence logical to investigate the number of degrees

5.2 Polarisation Resolution

of freedom within polarisation based systems, as will be determined by the size of the polarisation domain and the polarisation resolution of the PSA. In this vein the number of distinguishable states is defined here to be the ratio of the volume of uncertainty in polarisation space before a measurement to the volume of uncertainty after a measurement (and hence can equally be called a metric of fractional accuracy). Explicitly the number of degrees of freedom is defined as

$$\tilde{\mathcal{N}}_\mathcal{S} = \tilde{\mathcal{A}}_w = \frac{V_\mathcal{S}}{\tilde{V}_\mathbf{w}}, \qquad (5.10)$$

and

$$\tilde{\mathcal{N}}_\mathcal{P} = \tilde{\mathcal{A}}_u = \frac{V_\mathcal{P}}{\tilde{V}_\mathbf{u}}, \qquad (5.11)$$

when considering encoding in Stokes and (reduced) Poincaré space respectively. An intuitive analog to this definition can be found in optical imaging whereby the uncertainty in an object's position before a measurement is merely the field of view of the imaging system, whilst afterwards, assuming a diffraction limited system with circular aperture, is the area of the Airy disc. Similarly, the uncertainty before a polarimetric measurement is the entirety of the associated Hilbert space. The volume is thus easily calculable using the Lebesgue measure and is given by $V_\mathcal{S} = \frac{4\pi}{3}\mathcal{D}$ for Stokes space, where $\mathcal{D}$ is the dynamic range of the photodetectors and $V_\mathcal{P} = \frac{4\pi}{3}$ for Poincaré space i.e. if $N_u = 3$. If one or more of the polarisation parameters are known a priori only the volume of space spanned by the unknown parameters need be considered.

Finally, a local accuracy can also be defined analogously to Eqs. (5.10) and (5.11) if the FIM before Bayesian averaging, i.e. $\mathbb{J}_\mathbf{w}^r$ or $\mathbb{J}_\mathbf{u}^r$ is used to define the volume of the ellipsoid of concentration for a given incident polarisation state. The accuracy of a PSA is a useful concept when considering a novel optical data storage solution, for example, in which information is encoded into the polarisation state of light scattered from data pits orientated at different angles [79], since it dictates the storage capacity increase possible.

5.2.2 Efficiency of Observation

When performing experiments in conditions with limited light levels, for example in single molecule studies [25], it is important to utilise the detected photons as efficiently as possible. In the context of polarimetric experiments this implies achieving the greatest accuracy, or polarisation resolution, per photon. Physical limits however exist as to the extent to which this can be achieved. To establish these limits note that the reciprocal of the accuracy of a PSA can be considered as the fractional volume of uncertainty in polarisation space, or the probability of measuring a state of polarisation lying within the same volume, if all polarisation states were equally likely. One can thus (following Shannon [72]) associate an information gain from a polarimetric observation as the logarithm of the accuracy i.e.

$$\widetilde{\mathcal{I}} = -\log_2\left(\frac{\widetilde{V_u}}{V_P}\right) = \log_2 \widetilde{\mathcal{A}}. \quad (5.12)$$

The relationship between physical entropy in the thermodynamical sense and information has been known for many years and was first recognised by Szilard [77] and later applied by Brillioun [8–11]. The relationship states that information $\mathcal{I}$ about a system can *only* be obtained if there is an increase in entropy ΔH such that

$$\widetilde{\mathcal{I}} \leq \frac{\Delta H}{k_B \ln 2} \quad (5.13)$$

where $k_B = 1.381 \times 10^{-23}$ m^{2}kg s^{-1}K^{-1} is Boltzmann's constant. Equality is only achieved for a reversible observation. From inequality (5.13) it is thus possible to define the *efficiency of observation*, η, $(0 \leq \eta \leq 1)$

$$\eta = \frac{\widetilde{\mathcal{I}} k_B \ln 2}{\Delta H}. \quad (5.14)$$

Consider then a single optical detector which makes an observation by absorption of n_{i0} photons with mean energy $h\nu_0$.[4] The total energy absorbed will eventually be dissipated as heat corresponding to an increase in the entropy of the detector. The second law of thermodynamics then dictates that $\Delta H_i = n_{i0}h\nu_0/\Theta$, where Θ is the thermal noise temperature, i.e. ambient temperature, of the detector. There will however be an entropy cost for each measurement made such that total entropy cost is given by

$$\Delta H = \sum_{i=0}^{N_D-1} \Delta H_i + \Delta H_a, \quad (5.15)$$

where ΔH_a represents the entropy cost associated with photons that are not absorbed in the detectors. Ultimately these "lost" photons will also be absorbed by some material body at temperature Θ_0 and again dissipated as heat such that $\Delta H_a = (1 - \text{tr}[\mathbb{V}])s_0/\Theta_0$ for simultaneous measurements or alternatively $\Delta H_a = (N_D - \text{tr}[\mathbb{V}])s_0/\Theta_0$ for sequential observations. If all photons are absorbed by photodetectors (i.e. the PSA is 100% light efficient), the efficiency of observation is given by

---

[4] Entropy is an average property of a system and hence it is sufficient to consider the average number of photons absorbed, n_{i0}, as opposed to a particular realisation of the observation process in which n_i photons are absorbed.

5.2 Polarisation Resolution

$$\eta = \frac{k_B \Theta \ln 2}{n_0 h \nu_0} \log_2\left(\frac{V_p}{\widetilde{V}_\mathbf{u}}\right), \qquad (5.16)$$

where $s_0 = \sum_i n_{i0} h \nu_0 = n_0 h \nu_0$.

5.2.3 Examples

Accuracy, information and efficiency of observation have all been shown to be dependent on the FIM (which is averaged under the assumption of maximal ignorance to form the BFIM) and as such all that remains to quantify system performance in polarisation measurements is to calculate $\mathbb{J}_\mathbf{w}^r$ and $\mathbb{J}_\mathbf{w}$. This however requires making some assertions as to the type of noise present in the system and of the PSA configuration. In the following numerical calculations the reduced FIM (and BFIM) associated with estimation of $\mathbf{u}$ will only be considered so as to elucidate the polarisation dependent performance characteristics of different PSAs.

5.2.3.1 Noise Models and Fisher Information

In what follows the two noise regimes discussed in Chap. 2 will be considered; namely Poisson and Gaussian statistics. The first example discussed considers the quantisation of classical light, which produces Poisson distributed noise on the detector with variance n_{i0}, where $n_{i0} = E[n_i]$ is the mean number of photons absorbed by the ith detector. The second example meanwhile assumes that the mean intensity is large enough so as to invoke the Central Limit Theorem, however considers the improvement that can be achieved when using non-classical, squeezed light. Under these circumstances squeezed light produces Gaussian noise statistics with variance $s^2 n_0$ where $s^2 < 1$ is the squeezing factor [67]. The number of absorbed photons on a single detector is thus parameterised by the PDFs

$$f_{N_i}^{\text{sht}}(n_i|n_{i0}) = \frac{(n_{i0}+n_{ib})^{n_i}}{n_i!} \exp[-(n_{i0}+n_{ib})], \qquad (5.17)$$

$$f_{N_i}^{\text{sqz}}(n_i|n_{i0}) = \frac{1}{\sqrt{2\pi(s^2 n_{i0}+n_{ib})}} \exp\left[-\frac{(n_i - n_{i0} - n_{ib})^2}{2(s^2 n_{i0}+n_{ib})}\right], \qquad (5.18)$$

respectively.[5] $D_i = n_i h \nu_0$ is then the detected intensity on the ith detector.[6] The additional mean term n_{ib} has been introduced in Eqs. (5.17) and (5.18) to account for other potential additive sources of stray photons, assumed to be independent and Poisson (Gaussian) distributed, such that the joint PDF is also Poisson (Gaussian)

---

[5] It is assumed that the average number of photons is large enough that the probability of negative n_i is negligible in accordance with physical reality.
[6] Throughout this work the quantum efficiency is assumed to be unity for simplicity.

distributed (see discussion in Sect. 2.3.1). A good discussion of such possible noise sources is given in [6], however two simple examples would be a detector dark count or a passive background. Although not necessary, the simplifying assumption that these additional noise sources affect each detector equally such that $n_{ib} = n_b$ is also made. Furthermore, it is reasonable to assume that the noise present on each of the N_D measurements is independent and hence the joint PDF required for calculation of the FIM is given by $f_\mathbf{N}(\mathbf{n}|\mathbf{n}_0) = \prod_{i=1}^{N_D} f_{N_i}(n_i|n_{i0})$.

Using Eqs. (3.28), (5.17) and (5.18) the FIMs for polarisation measurements are given by

$$\mathbb{J}_\mathbf{w}^r = \mathbb{G}^T \mathbb{J}_\mathbf{D}^r \mathbb{G}, \tag{5.19}$$

where

$$\mathbb{J}_\mathbf{D}^r = \frac{1}{h^2 \nu_0^2} \operatorname{diag}\left[\frac{1}{n_{i0} + n_b}\right], \tag{5.20}$$

assuming classical shot noise and

$$\mathbb{J}_\mathbf{D}^r = \frac{1}{h^2 \nu_0^2} \operatorname{diag}\left[\frac{1}{s^2 n_{i0} + n_b}\right] \tag{5.21}$$

for Gaussian noise, whilst

$$\mathbb{G} = \frac{\partial \mathbf{D}}{\partial \mathbf{w}} = \mathbb{V}\mathbb{T}\frac{\partial \mathbf{S}}{\partial \mathbf{w}} \tag{5.22}$$

and $\partial \mathbf{S}/\partial \mathbf{w} = \operatorname{diag}[s_0, s_0, s_0, 1]$. It is immediately apparent from Eqs. (5.20) and (5.21) that use of squeezed light gives an improvement in performance over classical light. Although the potential performance gains from squeezed light have been previously reported in the context of imaging, e.g. [19, 67, 68, 80], this result has not previously been shown for polarimetric studies. It is interesting to note that $\mathbb{J}_\mathbf{D} \propto 1/\nu_0^2$. The increase in Fisher information (and associated increase in system accuracy) with lower frequencies arises since this corresponds to more collected photons for a given intensity which, as discussed in [5], corresponds to more independent samples of the stochastic variables. The BFIMs are then found by performing the Bayesian averaging of Eqs. (5.20) and (5.21).

5.2.3.2 Polarimeter Architectures

For definiteness these results are illustrated using three DOAPs existing in the literature. The first of these, as proposed by Azzam [2] and shown schematically in Fig. 5.2a, can be easily implemented using only beam splitters, polarisers and waveplates. The detectors in turn project the incident Stokes vector on to horizontal,

5.2 Polarisation Resolution

vertical, linear 45° and right circular polarised states, and hence has the instrument matrix

$$\mathbb{T}_1 = \frac{1}{2} \begin{pmatrix} 1 & 1 & 0 & 0 \\ 1 & -1 & 0 & 0 \\ 1 & 0 & 1 & 0 \\ 1 & 0 & 0 & 1 \end{pmatrix}. \tag{5.23}$$

Any noise in the intensity measurements is amplified during data processing to extract, for example, the Stokes parameters, the extent of which is often measured using the condition number of the associated matrices, ($\mathbb{T}$ for a Stokes polarimeter), defined as $\kappa_T = \|\mathbb{T}\|_F \|\mathbb{T}^{-1}\|_F$ where $\|\ldots\|_F$ denotes the Frobenius norm.[7] Compain and Drevillon [15] proposed an alternative PSA construction as shown in Fig. 5.2b in which the prism geometry and the angle of incidence of light onto the first surface are optimised to minimise the condition number of the instrument matrix to a value of 4.48. The associated instrument matrix is

$$\mathbb{T}_2 = \frac{1}{2} \begin{pmatrix} 1 & -0.575 & 0.818 & 0 \\ 1 & -0.575 & -0.818 & 0 \\ 1 & 0.617 & -0.003 & 0.787 \\ 1 & 0.617 & 0.003 & -0.787 \end{pmatrix}. \tag{5.24}$$

Note that the deviations of $\mathbb{V}$ from the ideal case arising from Fresnel reflection and transmission at the prism entrance surface will be ignored so that the results calculated will be comparable to the alternative DOAPs considered here. The imbalance between reflected and transmitted beams only equates to ≈5% however and discrepancies from reality will thus be small. Instrument matrices with smaller condition numbers than 4.48 are possible [70], however experimental realisation of these is complicated since it requires, in general, eight Babinent Soleil compensators.

Finally a DOAP configuration employing a basis of six distinct measurement states, as given by the instrument matrix

$$\mathbb{T}_3 = \frac{1}{2} \begin{pmatrix} 1 & 1 & 0 & 0 \\ 1 & -1 & 0 & 0 \\ 1 & 0 & 1 & 0 \\ 1 & 0 & -1 & 0 \\ 1 & 0 & 0 & 1 \\ 1 & 0 & 0 & -1 \end{pmatrix} \tag{5.25}$$

was recently proposed by Lara and Paterson [42] and is schematically shown in Fig. 5.2c. This DOAP architecture was shown to possess polarisation independent noise characteristics in the Stokes parameters as inferred from Eq. (5.2), in the presence of a combination of Gaussian thermal noise and signal dependent Poisson noise.

---

[7] The condition number is frequently defined using alternative matrix norms, however the Frobenius norm will be used throughout this chapter.

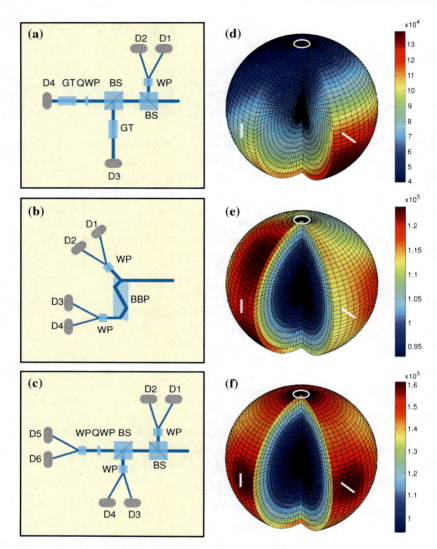

Fig. 5.2 *Left* schematics of three alternative DOAP designs (see text). Notation is as follows: *BS* beam splitter, *WP* Wollaston prism, *QWP* quarter wave plate, *GT* Glan Thompson polariser, *BBP* broadband prism and *D* detector. *Right* Poincaré diagrams showing the polarisation state dependence of accuracy, before Bayesian averaging for each DOAP configuration. Simulation parameters used were $n_0 = 10^4$, $n_b = 0$, $\lambda_0 = 405$ nm and $c = 0.9$. White markers denote the state of polarisation at different points on the Poincaré sphere and are shown for reference

Using Eqs. (5.8), and (5.19)–(5.25), $\mathbb{J}_{\mathbf{u}}^r$ can be calculated and hence so too can the accuracy before Bayesian averaging $\mathcal{A}_u = \mathcal{A}_u(\mathbf{u})$. Due to the similarity of the form of the FIM for Gaussian and Poisson noise, restriction is now made to Poisson noise

5.2 Polarisation Resolution

only. Numerical calculations were performed for each DOAP configuration assuming incident light with wavelength of 405 nm. A total mean photon count of 10^4 and a zero background count were further assumed. The resulting state dependent PSA accuracy is shown in Figs. 5.2d–f. Whilst these plots are formed via direct evaluation of the formulae given in the preceding theory, Monte-Carlo simulations (in which the accuracy was calculated from the covariance matrix of simulated random data) are in good agreement.

With reference to Fig. 5.2 and Eqs. (5.23)–(5.25) it is worth mentioning that, for a particular polarimeter architecture, the best accuracy is achieved when measuring totally polarised states that equalise the intensity measured in each polarimeter arm, a result which also holds for general PSA configurations. Accuracy is however seen to decrease with the degree of polarisation, i.e. toward the centre of the Poincaré sphere, a trend which would be expected.

It is also noted that the PSA architecture proposed by Lara and Paterson is not seen to give a constant accuracy over the surface of the Poincaré sphere, in apparent contradiction to [42]. This discrepancy however arises due to use of a different metric. The metric proposed in [42] is equivalent to tr$[\mathbb{J}^{-1}]$, or so-called A-optimality [48], however the metric proposed in this work takes greater account of the cross-correlations present in parameter estimates. The six arm DOAP does however still exhibit greater uniformity over the surface of the Poincaré sphere, resulting from a greater sampling of Poincaré space.

If it were known a priori that some particular polarisation state was more likely to be measured, additional accuracy gains could be made (as described by the $\mathbb{J}_\mathbf{w}^{ap}$ term of Eq. (3.46)), by appropriate system design (see Sect. 5.3 for a fuller discussion).

Figure 5.3 shows the variation of accuracy and efficiency of observation (after averaging) as a function of the number of absorbed photons for each DOAP. The accuracy is seen to improve as the number of detected photons increases. Infinite accuracy is thus, in principle, possible in polarimetry if enough photons are detected. A similar conclusion was reached in terms of localisation accuracy for two point objects [58, 64]. Additionally the efficiency falls as photon numbers increase. This essentially arises since there is a redundancy in the information which each photon in a beam carries with regard to their polarisation due to the correlations that exist between them.

Whilst a relatively low number of photons were considered when calculating the data in Fig. 5.3 (given the efficiency properties of the MLE described in Sect. 3.3.3) hence throwing the validity of the plots into question, it should be observed that both accuracy and efficiency of observation are monotonic functions of the mean photon count. Consequently trends inferred from Fig. 5.3 are valid for Poisson noise. Furthermore if Gaussian noise described by Eq. (5.18) was assumed in numerical calculations, plots of identical functional form would follow. Due to the exact efficiency of the MLE in Gaussian noise Fig. 5.3 is valid even at such low photon counts.

Some final reflections must be made in the context of quantum mechanics. Whilst a quantum analog to Fisher information can be formulated in terms of projection operators [7, 35, 36] and similar metrics formed, it is currently unclear (at least to the author) as to the role of the non-commutativity of the Stokes parameters for

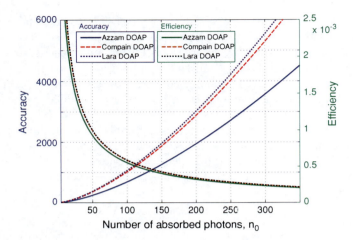

Fig. 5.3 Variation of accuracy and efficiency of observation with mean photon count for different DOAP configurations (see text). Simulation parameters are the same as Fig. 5.2

example. Work is ongoing to resolve these questions in the hope of generalising the definition of polarisation resolution to a quantum domain.

5.2.4 Channel Capacity and Detector Numbers

Using Frieden's definition of channel capacity it is possible to consider the performance characteristics of polarimeters as the number of detector arms is varied more rigorously. Consider first the channel capacity in the context of estimating the mean intensities only, i.e. $\text{tr}(\mathbb{J}_\mathbf{D})$. For a Stokes polarimeter, and assuming a Poisson noise regime, Eq. (5.20) gives the Fisher capacity as

$$C_\mathbf{D} = \sum_{i=1}^{N_D} \frac{1}{n_{i0} h^2 \nu_0^2} = \sum_{i=1}^{N_D} \frac{1}{h \nu_0 D_{i0}}, \tag{5.26}$$

which under the constraint $0 < \sum_{i=1}^{N_D} D_{i0} \leq s_0$ is a maximum when the intensity is equalised across the detectors, such that $D_{i0} = a s_0 / N_D$, where $a \leq 1$ is a positive constant (although henceforth assumed to be unity). Equalising the intensity on each detector also equalises the noise which is considered a desirable property in optimisation of polarimeters [82, 86]. This result will also be justified further in Sect. 5.3.1. The channel capacity is thus bounded according to[8]

---

[8] A factor of $h\nu_0$ shall henceforth be dropped for clarity.

5.2 Polarisation Resolution

$$C_\mathbf{D} \leq \frac{N_D^2}{s_0 + N_D D_b},\quad (5.27)$$

where $D_b = n_b h \nu_0$ is the mean background intensity reading. Although channel capacity initially increases quadratically with the number of detectors, it is seen this slows to a linear increase as the additive noise term D_b becomes more dominant on each detector. If a Gaussian noise model with constant covariance were used, the channel capacity would increase linearly with N_D. Since Fisher information is additive [27] the Fisher channel capacity for a Mueller polarimeter can be calculated by summing the capacities for each input polarisation state. Thus

$$C_\mathbf{D} \leq \frac{N_R N_D^2}{R_0 + N_D D_b},\quad (5.28)$$

where R_0 is the mean intensity of the probing polarisation states. Achieving this bound is unlikely though, since equality requires the sample to be perfectly transmitting to all incident polarisation states. Equation (5.28) shows however that the maximum channel capacity for a Mueller polarimeter only scales linearly with respect to the number of probing polarisation states.

Parameter inference problems however introduce derivative factors, of the form $\partial \mathbf{D}/\partial \mathbf{w}$, into the definition of the FIM. Since these derivative terms scale with the incident intensity, it can also be shown (assuming equal peak intensity on every detector, such that $D_{i0} = D_0$) that $C_\mathbf{w} = \text{tr}[\mathbb{J}_\mathbf{w}] \propto D_0 \propto s_0/N_D$.

Practically these results embody the intuitive result that by performing more measurements i.e. increasing the sampling, greater redundancy is introduced into the experimental data, hence allowing a better precision in parameter estimates to be achieved. A larger intensity per measurement is however preferable. Henceforth it will be assumed that $N_D = N_R = 4$ since this is the minimum number of measurements required to determine $\mathbf{S}$ or $\mathbb{M}$ uniquely in a noise free system. The associated FIMs are thus 4×4 and 16×16 respectively.

5.3 Optimisation of Polarimeters

Much effort has been invested into determining optimal configurations of polarimeters in terms of their experimental setup e.g. [1, 18, 32, 33, 76, 81, 83, 86], however invariably little consideration is given to a priori information that an experimenter may have about the system they are studying. Information theory however states, that if exploited correctly, such information can improve the accuracy of any measurements [49, 66]. For example if the position of an object is approximately known, the field of view can be reduced, perhaps by using a confocal microscope, giving rise to an increase in the bandwidth and hence resolution of the system [17]. To address this omission consideration is now given to how such a priori information can be represented and incorporated into system optimisation, as fully detailed in [24].

In doing so, it is found that common optimisation procedures do not necessarily give the optimal polarimeter configuration. This point is perhaps most easily illustrated by restricting to a Poisson noise model for the remainder of this chapter. Furthermore, the optimisation framework naturally describes the distribution of errors amongst inferred polarisation parameters, such as diattenuation, retardance and depolarisation as may be obtained from a Lu–Chipman Mueller matrix decomposition [45]. These results could potentially be used for a more accurate noise analysis in polarimetry.

An experimenter may know from an existing model or earlier data that the object being studied belongs to a restricted class, that is to say they possess some a priori information about the parameters being measured. Known restrictions on the possible values of **w** can often be conveniently parameterised using a PDF $f_\mathbf{W}(\mathbf{w})$ which describes the probability of each value of **w** occurring.

Optimisation of both Stokes and Mueller polarimeters will be considered in the presence of such a priori information and will be performed in terms of the full polarisation resolution, as parameterised by $|\mathbb{J}_\mathbf{w}|$, by varying the instrument matrix, beam splitting ratios and, in the case of Mueller polarimeters, the incident polarisation states. In particular, when considering Stokes polarimeters the parameter vector is set to $\mathbf{w} = \mathbf{S}$ or alternatively $\mathbf{M} = \text{vec}[\mathbb{M}]$ for Mueller polarimeters, where $\text{vec}[\cdots]$ denotes the vectorisation (or stacking) operation. Optimisation in terms of the Stokes vector and vectorised Mueller matrix is sufficient since the $\partial \mathbf{S}/\partial \mathbf{w}$ factor of Eq. (5.22) does not contain any free design parameters in either type of polarimetry.

Setting $\mathbf{w} = \mathbf{S}$ or $\mathbf{M}$ and using Eqs. (3.28), (5.1) and (5.4) it is possible to calculate the relevant BFIMs as

$$\mathbb{J}_\mathbf{S} = \mathbb{T}^T \mathbb{V}^T \mathbb{J}_\mathbf{D} \mathbb{V} \mathbb{T}, \tag{5.29a}$$

$$\mathbb{J}_\mathbf{M} = \left(\mathbb{R} \otimes \mathbb{V}^T \mathbb{T}^T \right) \mathbb{J}_\mathbf{D} \left(\mathbb{R}^T \otimes \mathbb{T} \mathbb{V} \right), \tag{5.29b}$$

where $\mathbb{J}_\mathbf{D} = E[\mathbb{J}_\mathbf{D}^r] + \mathbb{J}_\mathbf{D}^{ap}$ (c.f. Eq. (3.46)), $\mathbb{J}_\mathbf{D}^r$ is of the form of Eq. (5.20) and $\otimes$ denotes the Kronecker product. Hence,

$$|\mathbb{J}_\mathbf{S}| = |\mathbb{V}|^2 |\mathbb{T}|^2 |\mathbb{J}_\mathbf{D}|, \tag{5.30a}$$

$$|\mathbb{J}_\mathbf{M}| = |\mathbb{R}|^8 |\mathbb{V}|^8 |\mathbb{T}|^8 |\mathbb{J}_\mathbf{D}|. \tag{5.30b}$$

Closer inspection of Eqs. (5.30) reveals there are two factors which influence the amount of information received in an optical experiment. The first of these corresponds to the amount of information acquired during the physical measurement as described by the $|\mathbb{J}_\mathbf{D}|$ term. This component also encompasses any a priori information that may be possessed. The second, and perhaps the more familiar, influence is related to any subsequent data processing used to extract the Stokes vector or Mueller matrix from the measured intensities as per Eqs. (5.2) and (5.5). As previously mentioned such data processing is often parameterised by the condition number of the associated matrices however since the condition number of a matrix is inversely

5.3 Optimisation of Polarimeters

proportional to its determinant (see e.g. [70]) the noise amplification is equally described by the $|\mathbb{V}|$, $|\mathbb{T}|$ (and $|\mathbb{R}|$) factors of Eqs. (5.30).

Frequently the condition number of the instrument matrix (and input polarisation matrix) is used as a figure of merit for polarimeter optimisation [1, 18, 76, 83]. The discussion above however has highlighted the inadequacy of this strategy, in general, since it gives no regard to potential gains that can be made by improving the precision of the measurement itself or incorporation of a priori knowledge. Use of the informational figure of merit proposed here is hence more holistic in terms of measuring the quality of a polarimeter.

5.3.1 Examples

Having introduced a more suitable framework within which both Stokes and Mueller polarimeters can be optimised a number of examples are now given to highlight some points of interest. Circumstances under which optimisation in terms of polarisation resolution is equivalent to use of the condition number are highlighted in the first example, however further examples then illustrate that when a priori information is introduced this equivalence does not hold.

5.3.1.1 Maximal Ignorance

The first example continues the assumption made thus far of maximal ignorance of the likely incident polarisation states. Initially considering a Stokes polarimeter, the a priori information (or lack thereof) is again modelled by assuming each polarisation state is equally likely, however it is assumed that all possible incident states have the same intensity S_0 and degree of polarisation P.[9]

Since a uniform prior PDF implies $\mathbb{J}_\mathbf{D}^{ap} = 0$,

$$|\mathbb{J}_\mathbf{S}| = |\mathbb{V}|^2 |\mathbb{T}|^2 \prod_{i=1}^{4} E_\mathbf{S}\left[\frac{1}{D_{i0} + D_b}\right], \qquad (5.31)$$

where it can be shown

$$E_\mathbf{S}\left[\frac{1}{D_{i0} + D_b}\right] = \frac{1}{S_0 P} \log\left[\frac{(1+P)S_0 + 2D_b}{(1-P)S_0 + 2D_b}\right],$$
$$= \frac{2}{S_0 P} \operatorname{arctanh}\left[\frac{S_0 P}{S_0 + 2D_b}\right]. \qquad (5.32)$$

---

[9] PDFs in which P can vary can theoretically be used, however care must be taken since depolarisation can be introduced by the measurement instrumentation [53].

Hence

$$|\mathbb{J_S}| = \frac{1}{8}|\mathbb{V}|^2|\mathbb{T}|^2 \left(\operatorname{arctanh}\left[\frac{S_0 P}{S_0 + 2D_b}\right] \bigg/ S_0 P\right)^4. \tag{5.33}$$

It is thus apparent that to maximise the information obtained we must maximise $|\mathbb{V}|$ and $|\mathbb{T}|$ or equivalently make the associated condition number as small as possible. Considering the implications of minimisation of the condition number of $\mathbb{V}$ first (or alternatively maximising its determinant), it is noted that the determinant of a matrix with a fixed trace is maximised when it is diagonal with equal elements [63]. Consequently it can be concluded that DOAPs perform optimally when the intensity in each detector is equal. Sequential measurements automatically satisfy this condition since in the ideal case $\mathbb{V} = \mathbb{I}$. Improved performance from intensity equalisation has already been seen in Fig. 5.2 when considering the accuracy of the different PSA configurations, hence further supporting the conclusions made and also justifying the assumptions made in Sect. 5.2.4. Van der Sluis detailed an equalising theorem to reduce the condition number of a matrix [84], which has since been applied in the context of realistic, experimental polarimeter instrument matrices [47]. This equalising theorem exactly acts so as to achieve an equal intensity in each detector.

Minimising the condition number of the instrument matrix can be shown geometrically to correspond to maximising the volume of the tetrahedron whose vertices on the Poincaré sphere are defined by $\mathbf{T}_i$, i.e. making the tetrahedron regular [70]. Although the same conclusion has been previously reached via considerations of the structure of the instrument matrix and noise propagation [4, 70, 82, 83] the derivation presented here based on information theory appears to be new. Since a maximum determinant corresponds to minimal noise amplification the signal to noise ratio (SNR), given by

$$\operatorname{SNR} = \left(\frac{S_0}{2P} \operatorname{arctanh}\left[\frac{S_0 P}{S_0 + 2D_b}\right]\right)^{1/2}, \tag{5.34}$$

is also maximum. If $D_b \ll S_0$ this reduces to the familiar $S_0^{1/2}$ scaling associated with Poisson noise [19].

There are an infinite number of possible instrument matrices corresponding to the rotation of the tetrahedron within the Poincaré sphere about the origin, however given one optimal instrument matrix (as can easily be found numerically), e.g. [70]

$$\mathbf{T}_{\text{opt}} = \begin{pmatrix} 1 & 1 & 0 & 0 \\ 1 & -0.333 & -0.816 & 0.471 \\ 1 & -0.333 & 0 & -0.943 \\ 1 & -0.333 & 0.816 & 0.471 \end{pmatrix}, \tag{5.35}$$

it is possible to find alternative configurations by applying a suitable rotation matrix.

5.3 Optimisation of Polarimeters

For a Mueller matrix polarimeter

$$|\mathbb{J_M}| = |\mathbb{R}|^8 |\mathbb{V}|^8 |\mathbb{T}|^8 \prod_{i=1}^{16} E_\mathbf{M} \left[\frac{1}{D_{i0} + D_b} \right], \quad (5.36)$$

whereby the same results apply, however the determinant of $\mathbb{R}$ must also be maximised, that is to say the incident polarisation states must be made as orthogonal as possible for optimal performance. For the case of maximal ignorance there is no relationship between $\mathbb{T}$ and $\mathbb{R}$.

To conclude this example the Fisher channel capacity is again considered. From Eq. (5.32) it can be shown that

$$C_\mathbf{D} \propto \frac{1}{S_0 P} \text{arctanh} \left[\frac{S_0 P}{S_0 + 2D_b} \right]. \quad (5.37)$$

For light with a low degree of polarisation, whereby $P \ll 1$, the channel capacity approximately obeys

$$C_\mathbf{D} \propto \frac{1}{S_0 + 2D_b} + \frac{1}{3} \frac{S_0^2}{(S_0 + 2D_b)^3} P^2, \quad (5.38)$$

whilst for highly polarised light

$$C_\mathbf{D} \propto \frac{1}{S_0} \text{arctanh} \left[\frac{S_0 P}{S_0 + 2D_b} \right]. \quad (5.39)$$

Channel capacity thus increases with the degree of polarisation of light, as would be expected (see Fig. 5.4).

5.3.1.2 Matched Filter

Adopting the opposite extreme to maximal ignorance, consider now the polarimetric equivalent to the matched filter. Matched filters frequently arise in signal processing, for example in radar detection, in which a *known* signal (or template) is correlated with a measured one so as to determine the presence or absence of the known signal against some background. Furthermore the magnitude of the correlation can also be used to infer parameters of interest, such as the distance to objects. In this paradigm matched filters are designed so as to maximise the SNR when the template is present [85]. A similar approach is adopted here, however discussion is restricted to a Stokes polarimeter in which depolarised light constitutes the background signal, whilst the magnitude of a known state of polarisation is to be estimated.

Denoting the known polarisation state by its Stokes vector $\mathbf{S}_t = (S_{t0}, S_{t1}, S_{t2}, S_{t3})$ the a priori knowledge can be represented by the PDF $f(\mathbf{s}) = \delta(\mathbf{s} - \mathbf{S}_t)$ where $\delta(\mathbf{x})$ is

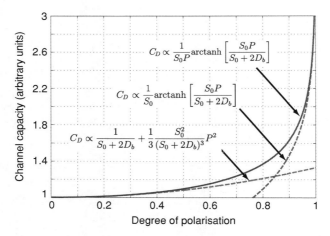

Fig. 5.4 Dependence of channel capacity on degree of polarisation for $S_0/D_b = 10^4$ (*solid*). Limiting cases for $P \ll 1$ and $P \approx 1$ are also shown (*dashed*)

the multi-dimensional Dirac delta function. Since **s** is non-random $\mathbb{J}_\mathbf{D}^{ap}$ is identically zero. Consequently $(\mathbb{J}_\mathbf{D})_{ij} = \delta_{ij}/(D_{i0} + D_b)$ whereby

$$|\mathbb{J}_\mathbf{S}| = |\mathbb{V}|^2 |\mathbb{T}|^2 \prod_{i=1}^{4} \frac{1}{D_{i0} + D_b}. \qquad (5.40)$$

If the additive noise term D_b is zero infinite information can be obtained if $D_{i0} = 0$ on a single detector, corresponding to one arm of the polarimeter projecting the incident polarisation on to the basis state $\mathbf{T}_i \propto (S_{t0}, -S_{t1}, -S_{t2}, -S_{t3})$. Note the parallel with conventional matched filters, for which the filter corresponds to the template reversed in time. This result can be understood by noting that for a given state of polarisation there are only two PSA configurations capable of uniquely identifying that state, namely $\mathbf{T}_i \propto (S_{t0}, \pm S_{t1}, \pm S_{t2}, \pm S_{t3})$ corresponding to diametrically opposite points on the Poincaré sphere. For example only a horizontal or vertical polariser can unambiguously identify horizontally polarised light (giving a maximum or null intensity respectively). When taking a single measurement, it is not in general possible to know which intensity level corresponds to the maximum, whilst a null intensity is more clearly identifiable. Furthermore, an underlying Poisson process has been assumed in which noise variations grow as the intensity grows and hence lower intensities give a better accuracy.

If present, a depolarised background necessitates a second, distinct polarimeter arm and also results in finite information. The situation is similar for non-zero D_b. Additional polarimeter arms improve estimation precision as discussed earlier in Sect. 5.2.4. Although unnecessary, the assertion that $N_D = 4$ is maintained to allow easy comparison. Once more it is found that there are an infinite number of possible

5.3 Optimisation of Polarimeters

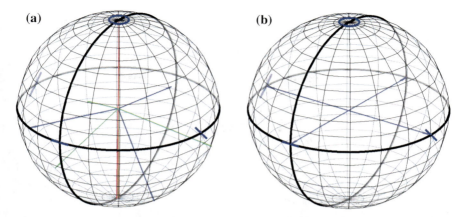

Fig. 5.5 Poincaré sphere showing possible polarimeter configurations (**a**) for a Stokes polarimeter matched to the template Stokes vector $(1, 0, 0, 1)$ i.e. right circularly polarised light and (**b**) a linear polarimeter, assuming the ratio $S_0/D_b = 10^5$. Each arrow denotes the basis Stokes vector of a polarimeter arm

instrument matrices that give rise to a maximum in the information, since it is possible to trade off precision in the intensity measurements (corresponding to higher light levels) with a reduction in the noise amplification associated with data processing i.e. smaller condition number. Three possible polarimeter configurations are shown in Fig. 5.5a for a template Stokes vector of $(1, 0, 0, 1)$. The first configuration (shown in green) gives the best condition number possible for a matched polarimeter (and consequently worse measurement accuracy), whilst the second (red) shows the opposite case, whereby the volume of the inscribed tetrahedron is significantly smaller i.e. larger condition number, yet the precision of measurement is increased since the total detected intensity is smaller. Practically, this arrangement is unsuitable since it is highly sensitive to alignment errors in the PSA. The third configuration (blue) illustrates a more general arrangement.

5.3.1.3 Linear Polarimeter

The final example assumes the polarisation incident into a Stokes polarimeter is restricted to lie on the equator of the Poincaré sphere. This could for example correspond to studying the light from a family of polarisers. Such a model could be useful in polarisation multiplexed optical data storage [53, 79] (see Chap. 7). Using the PDF $f(\varepsilon, \vartheta) = \delta(\varepsilon)/\pi$, where ε and ϑ are angles which define the position on the Poincaré sphere, the expectations can be evaluated analytically to give

$$E\left[\frac{1}{D_{i0} + D_b}\right] = \frac{1}{\sqrt{(S_0 + 2D_b)^2 + S_0^2 P^2 \cos^2 2\alpha_i}}, \quad (5.41)$$

where α_i is the equatorial angle on the Poincaré sphere for the ith basis Stokes vector of the instrument matrix. Considering it is known a priori that the incident polarisation state is linearly polarised there is no need to estimate S_3 since this only describes the ellipticity of the light and it can hence be treated as a nuisance parameter. A linear Stokes polarimeter is thus optimised when

$$|\mathbb{J}_\mathbf{u}| = |\mathbb{V}|^2 |\mathbb{T}|^2 \frac{\prod_{i=1}^{4} \left[(S_0 + 2D_b)^2 + S_0^2 P^2 \cos^2 2\alpha_i \right]^{-1/2}}{\sum_{i=1}^{4} \sin^2(2\alpha_i) \left[(S_0 + 2D_b)^2 + S_0^2 P^2 \cos^2 2\alpha_i \right]^{-1/2}}, \quad (5.42)$$

is maximum, where $\mathbf{u}$ denotes the parameter vector (S_0, S_1, S_2). In agreement with [81] the maximum of this metric occurs when the measurement basis Stokes vectors $\mathbf{T}_i$ are equally spaced around the equator of the Poincaré sphere as shown in Fig. 5.5b. When applied to Mueller polarimeters a similar analysis shows the optimal input polarisation states should also be equally spaced about the equator of the Poincaré sphere, although their position need bear no resemblance to those defined by $\mathbf{T}_i$.

5.3.2 Extension of Optimisation Results

The above results with regards to the optimisation of polarimeters hold not only for inference of the Stokes parameters or elements of a Mueller matrix, but can be further extended for inference of further polarisation parameters $\mathbf{z}$ derived from these quantities. Such a situation may arise, for example when performing a Lu–Chipman polar decomposition [45] on a measured Mueller matrix as is heavily used in the literature [12, 23, 43]. Equations (5.29) generalise to

$$\mathbb{J}_\mathbf{z} = \frac{\partial \mathbf{S}^T}{\partial \mathbf{z}} \mathbb{T}^T \mathbb{V}^T \mathbb{J}_\mathbf{D} \mathbb{V} \mathbb{T} \frac{\partial \mathbf{S}}{\partial \mathbf{z}}, \quad (5.43a)$$

$$\mathbb{J}_\mathbf{z} = \frac{\partial \mathbf{M}^T}{\partial \mathbf{z}} \left(\mathbb{R} \otimes \mathbb{V}^T \mathbb{T}^T \right) \mathbb{J}_\mathbf{D} \left(\mathbb{R}^T \otimes \mathbb{T} \mathbb{V} \right) \frac{\partial \mathbf{M}}{\partial \mathbf{z}}. \quad (5.43b)$$

Accordingly, the volume of the ellipsoid of concentration as found from $|\mathbb{J}_\mathbf{z}|$ is modified by a factor of $|\partial \mathbf{w}/\partial \mathbf{z}|^2$ ($\mathbf{w} = \mathbf{S}$ or $\mathbf{M}$), which is independent of $\mathbb{V}$, $\mathbb{T}$ and $\mathbb{R}$. Its significance in terms of optimisation with respect to the experimental setup is thus null, hence optimisation of these more complicated inference problems reduces to the optimisation procedure previously discussed.

5.4 Noise Propagation in Lu–Chipman Decomposition

5.4.1 Single Element Systems

Noise propagation in inference problems, that is to say how noise in experimental data manifests itself as errors in the parameters of interest, can also be considered by employing Eqs. (5.43). Although the mathematics is generally complicated, a result pertaining to polar decomposition of Mueller matrices [45] is explicitly given here. Before considering the composite systems for which polar decomposition is relevant a description of noise propagation for single polarisation element systems, namely pure diattenuators, retarders and depolarisers, must first be given.

A diattenuator is a non-depolarising optical element which preferentially transmits particular states of polarisation and has a Mueller matrix of the form [45]

$$\mathbb{M}_A = T_u \begin{pmatrix} 1 & \mathbf{A}^T \\ \mathbf{A} & \mathbf{m}_A \end{pmatrix}, \tag{5.44}$$

where T_u is the transmittance for unpolarised light, $\mathbf{A} = (A_1, A_2, A_3)$ is the diattenuation vector whose magnitude A is known as the diattenuation and

$$\mathbf{m}_A = \sqrt{1 - A^2}\,\mathbb{I} + (1 - \sqrt{1 - A^2})\frac{\mathbf{A}\mathbf{A}^T}{A^2}. \tag{5.45}$$

Any decomposition algorithm will need to estimate all four unknown parameters $\mathbf{z} = (A_1, A_2, A_3, T_u)$. Lengthy calculations give the derivatives required for evaluation of Eq. (5.43b) as

$$\frac{\partial \mathbb{M}_A}{\partial A_k} = T_u \begin{pmatrix} 0 & \boldsymbol{\delta}_k^T \\ \boldsymbol{\delta}_k & \frac{\partial \mathbf{m}_A}{\partial A_k} \end{pmatrix}, \qquad \frac{\partial \mathbb{M}_A}{\partial T_u} = \frac{\mathbb{M}_A}{T_u}, \tag{5.46}$$

where $k = 1, 2$ or 3, $\boldsymbol{\delta}_k = (\delta_{1k}, \delta_{2k}, \delta_{3k})^T$ and

$$\frac{\partial m_{Aij}}{\partial A_k} = \left[\frac{A_i \delta_{jk}}{A^2} + \frac{A_j \delta_{ik}}{A^2}\right]\left[1 - \sqrt{1 - A^2}\right]$$
$$+ \frac{A_k \delta_{ij}}{\sqrt{1 - A^2}} - \frac{A_i A_j A_k}{A^4}\left[2 + \frac{A^2 - 2}{\sqrt{1 - A^2}}\right], \tag{5.47}$$

where m_{Aij} is the (i, j)th element of $\mathbf{m}_A$. Using the FIM, $\mathbb{J}_\mathbf{A}$, as calculated from Eqs. (5.43)–(5.47), and the CRLB, the best obtainable precision for estimation of the diattenuation parameters can be calculated. The error on each parameter will in general be different, a point considered further in [54, 71].

Similarly consider a pure retarder which has a Mueller matrix of the general form

$$\mathbb{M}_R = \begin{pmatrix} 1 & \mathbf{0}^T \\ \mathbf{0} & \mathbf{m}_R \end{pmatrix}, \tag{5.48}$$

where

$$m_{Rij} = \delta_{ij}\cos R + \frac{R_i R_j}{R^2}(1-\cos R) + \sum_{q=1}^{3}\epsilon_{ijq}\frac{R_q}{R}\sin R. \tag{5.49}$$

Again (R_1, R_2, R_3) defines a retardance axis and has a norm of R, known as the retardance and ϵ_{ijq} is the Levi-Civita permutation symbol. Calculation of the FIM, $\mathbb{J}_\mathbf{R}$, requires the derivatives

$$\begin{aligned}\frac{\partial m_{Rij}}{\partial R_k} &= \left[\frac{R_i\delta_{jk}}{R^2} + \frac{R_j\delta_{ik}}{R^2}\right](1-\cos R) - \frac{R_k\delta_{ij}}{R}\sin R \\ &\quad - \frac{R_i R_j R_k}{R^2}\left[\frac{1-\cos R}{R} + \sin R\right] \\ &\quad + \sum_{q=1}^{3}\frac{\epsilon_{ijq}}{R^2}\left[R_q R_k \cos R + \left(R\delta_{qk} - \frac{R_k}{R}\right)\sin R\right].\end{aligned} \tag{5.50}$$

The case of a depolariser is however much more difficult to tackle since in general an eigen-analysis of the system is required to find the pertinent depolarisation parameters. This can not be described analytically except in some special cases. For example, if it were known a priori that the sample were a pure depolariser with Mueller matrix of the form

$$\mathbb{M}_\Delta = \begin{pmatrix} 1 & 0 & 0 & 0 \\ 0 & a & 0 & 0 \\ 0 & 0 & b & 0 \\ 0 & 0 & 0 & c \end{pmatrix}, \quad |a|, |b|, |c| \leq 1, \tag{5.51}$$

where $1-|a|$, $1-|b|$ and $1-|c|$ are the principal depolarisation factors, the derivatives

$$\frac{\partial M_{\Delta ij}}{\partial a} = \delta_{i2}\delta_{j2}, \quad \frac{\partial M_{\Delta ij}}{\partial b} = \delta_{i3}\delta_{j3}, \quad \frac{\partial M_{\Delta ij}}{\partial c} = \delta_{i4}\delta_{j4}, \tag{5.52}$$

can be easily calculated where $M_{\Delta ij}$ is the (i, j)th element of $\mathbb{M}_\Delta$. The appropriate FIM $\mathbb{J}_\Delta$ is then given by substituting Eq. (5.52) into Eq. (5.43b).

5.4.2 Composite Systems

The single element results discussed above can be used for noise analysis when the experimenter has a priori knowledge about the structure of the Mueller matrix.

5.4 Noise Propagation in Lu–Chipman Decomposition

If however this is not the case a Lu–Chipman decomposition is frequently performed so as to parameterise the sample. Fundamental to the Lu–Chipman decomposition is the fact that an arbitrary Mueller matrix can be written as the product of three distinct Mueller matrices corresponding to a depolariser, retarder and diattenuator i.e. $\mathbb{M} = \mathbb{M}_\Delta \mathbb{M}_R \mathbb{M}_A$. Morio and Goudail [52] considered the importance of altering the order in which the product is evaluated and found that different decompositions gave either unphysical results or merely comprised of an appropriate rotation compared to the Lu–Chipman decomposition and thus is only a mathematical, not physical, difference. With these results in mind the original formulation is adhered to, since this ensures physicality and furthermore corresponds to common usage. Alternative Mueller matrix decompositions are also starting to emerge in the field [16, 60, 69], however these will not be considered here since they have, as yet, not seen widespread use.

Calculation of the FIM for a Lu–Chipman decomposition can be achieved by application of the product rule to Eq. (5.43b) which yields

$$\mathbb{J}_z = \frac{\partial \mathbb{M}_A^T}{\partial z} \left(\mathbb{R} \otimes \mathbb{M}_R^T \mathbb{M}_\Delta^T \mathbb{T}^T \mathbb{V}^T \right) \mathbb{J}_D \left(\mathbb{R}^T \otimes \mathbb{V} \mathbb{T} \mathbb{M}_\Delta \mathbb{M}_R \right) \frac{\partial \mathbb{M}_A}{\partial z}$$
$$+ \frac{\partial \mathbb{M}_R^T}{\partial z} \left(\mathbb{M}_A \mathbb{R} \otimes \mathbb{M}_\Delta^T \mathbb{T}^T \mathbb{V}^T \right) \mathbb{J}_D \left(\mathbb{R}^T \mathbb{M}_A^T \otimes \mathbb{V} \mathbb{T} \mathbb{M}_\Delta \right) \frac{\partial \mathbb{M}_R}{\partial z}$$
$$+ \frac{\partial \mathbb{M}_\Delta^T}{\partial z} \left(\mathbb{M}_R \mathbb{M}_A \mathbb{R} \otimes \mathbb{T}^T \mathbb{V}^T \right) \mathbb{J}_D \left(\mathbb{R}^T \mathbb{M}_A^T \mathbb{M}_R^T \otimes \mathbb{V} \mathbb{T} \right) \frac{\partial \mathbb{M}_\Delta}{\partial z}, \quad (5.53)$$

since the structure of the matrices dictates that the cross terms are identically zero. It is important to note that the parameters of interest have been stacked into a single parameter vector for example $\mathbf{z} = (R_1, R_2, \ldots, b, c)$. $\mathbb{J}_z$ is then block diagonal

$$\mathbb{J}_z = \begin{pmatrix} \mathbb{J}_R & \mathbb{O} & \mathbb{O} \\ \mathbb{O} & \mathbb{J}_A & \mathbb{O} \\ \mathbb{O} & \mathbb{O} & \mathbb{J}_\Delta \end{pmatrix}, \quad (5.54)$$

where the order of the diagonal terms depends only on the ordering of the parameters in $\mathbf{z}$. Mathematically, the FIMs $\mathbb{J}_R$, $\mathbb{J}_A$ and $\mathbb{J}_\Delta$ are of the same form as the single element FIMs described in the previous section albeit for a slight modification in the effective input polarisation states and instrument matrix respectively, as can be seen by comparing Eqs. (5.43b) and (5.53). Fortunately this makes physical sense considering the Mueller matrix polar decomposition models the system as a cascade of three independent polarisation elements. Once more, it is important to give the cautionary note that the derivatives required to calculate $\mathbb{J}_\Delta$ can not be found analytically in general.

5.5 Conclusions

The work undertaken in this chapter set out to consider the informational limits in both Stokes and Mueller polarimetry. In this vein, and in analogy to definitions in other fields of research, a polarisation resolution was defined by employing the concepts of Fisher information introduced in Chap. 3. Both local and Bayesian definitions were given, where the latter considers the potential random nature of the measurement and experimental process. Further figures of merit were derived from the given definition of polarisation resolution. In particular, the number of degrees of freedom and the efficiency of observation were defined, which may be more pertinent measures in polarisation multiplexed systems, or low light experiments respectively.

By consideration of a number of existing polarimeter architectures it was also demonstrated that infinite accuracy in polarisation space is possible, provided enough photons are detected, with squeezed light achieving this more efficiently than classical light. The results given can be regarded as fundamental limits, since the noise models considered arise from the nature of light itself and not from external sources. It should however be noted that the results given in this chapter can only be considered to hold in a statistical sense. It is entirely feasible that better performance is achievable in a single instance, however if this is the case then it is merely a case of good fortune.

In the second half of this chapter the proposed definition of polarisation resolution was used to determine the optimal polarimeter configuration initially under the assumption that no knowledge was possessed as to the likely states of polarisation (or the Mueller matrix being measured). In so doing, it was found that polarisation resolution gives a holistic approach to optimisation, by automatically incorporating noise amplification, raw limits in the photodectection and signal equalisation among multiple measurements.

Additionally, the question as to how a priori information that may be possessed about a system under study can be used to improve the precision of measurements was addressed. Under these circumstances it was found that frequently used optimisation routines, in which the condition number of the instrument (and incident polarisation) matrix is maximised, are unsatisfactory and do not give optimal results. This was illustrated by considering the polarimetric equivalent of a matched filter and linear polarimeters. Specifically it was found that, under conditions in which the variance of the noise increases with incident intensity (such as Poisson noise or some types of Gaussian noise) optimal polarimeters are such that the measured intensity is jointly equalised among each detector. Fully polarised states of light were also seen to be measurable with greater resolution than partially polarised states. Although formulated within rather specific noise and estimation problems, optimisation with respect to Fisher information is easily extended to different regimes and is thus applicable to a wide variety of optical experiments, even outside the domain of polarimetry.

Inference problems in polarimetry present further cause for consideration, since noise propagation will not be balanced among each inferred parameter. This fact was highlighted by extending the definition of polarisation resolution to calculation of the FIM pertaining to a Lu–Chipman polar decomposition of a Mueller matrix, due to its widespread use in polarimetric analysis.

References

1. A. Ambirajan, D.C. Look, Optimum angles for a polarimeter. Opt. Eng. **34**, 1651–1658 (1995)
2. R.M.A. Azzam, N.M. Bashara, Division-of-amplitude photopolarimeter (DOAP) for the simultaneous measurement of all four Stokes parameters of light. J. Mod. Opt. **29**, 685–689 (1982)
3. R.M.A. Azzam, N.M. Bashara, *Ellipsometry and Polarised Light* (North Holland, Amsterdam, 1987)
4. R.M.A. Azzam, F.F. Sudradjat, Single-layer-coated beam splitters for the division-of-amplitude photopolarimeter. Appl. Opt. **44**, 190–196 (2005)
5. H.H. Barrett, J.L. Denny, R.F. Wagner, K.J. Myers, Objective assessment of image quality. II: Fisher information, Fourier crosstalk, and figures of merit for task performance. J. Opt. Soc. Am. A **12**, 834–852 (1995)
6. A. Bénière, F. Goudail, M. Alouini, D. Dolfi, Estimation precision of degree of polarization in the presence of signal-dependent and additive Poisson noises. J. Eur. Opt. Soc. Rap. Publ. **3**, 08002 (2008)
7. S.L. Braunstein, Quantum limits in precision measurement of phase. Phys. Rev. Lett. **69**, 3598–3601 (1992)
8. L. Brillioun, Maxwell's demon cannot operate: information and entropy I. J. Appl. Phys. **22**, 334–337 (1950)
9. L. Brillioun, Physical entropy and information II. J. Appl. Phys. **22**, 338–343 (1950)
10. L. Brillioun, The negentropy principle of information. J. Appl. Phys **24**, 1152–1163 (1953)
11. L. Brillioun, *Science and Information Theory* (Academic Press Inc., New York, 1956)
12. J.M. Bueno, Depolarization effects in the human eye. Vis. Res. **41**, 2687–2696 (2001)
13. R.A. Chipman, *Handbook of Optics*, vol 2 (McGraw Hill, New York, 1995)
14. E. Collett, Automatic determination of the polarization state of nanosecond laser pulses. U.S. Patent 4158506, 1979
15. E. Compain, B. Drevillon, Broadband division-of-amplitude-polarimeter based on uncoated prisms. Appl. Opt. **37**, 5938–5944 (1998)
16. J.M. Correas, P.A. Melero, J.J. Gil, Decomposition of Mueller matrices into pure optical media. Monografás del Seminario Matemaático Garcá de Galdeano **27**, 233–240 (2003)
17. I.J. Cox, C.J.R. Sheppard, Information capacity and resolution in an optical system. J. Opt. Soc. Am. A **3**(8), 1152–1158 (1986)
18. A. De Martino, E. Garcia-Caurel, B. Laude, B. Drévillon, General methods for optimized design and calibration of Mueller polarimeters. Thin Solid Films **455–456**, 112–119 (2004)
19. V. Delaubert, N. Treps, C. Fabre, A. Maître, H.A. Bachor, P. Réfrégier, Quantum limits in image processing. Europhys. Lett. **81**, 44001 (2008)
20. J. Ellis, A. Dogariu, Optical polarimetry of random fields. Phys. Rev. Lett. **95**, 203905 (2005)
21. S.G. Evangelides, L.F. Mollenauer, J.P. Gordon, N.S. Bergano, Polarization multiplexing with solitons. J. Lightwave Technol. **10**, 28–35 (1992)
22. P.B. Fellgett, E.H. Linfoot, On the assessment of optical images. Philos. Trans. Roy. Soc. Lond. A **247**, 369–407 (1955)
23. M. Floc'h, G. Le Brun, J. Cariou, J. Lotrian, Experimental characterization of immersed targets by polar decomposition of the Mueller matrices. Eur. Phys. J. Appl. Phys. **3**, 349–358 (1998)
24. M.R. Foreman, C. Macías Romero, P. Török, A priori information and optimisation in polarimetry. Opt. Express **16**, 15212–15227 (2008)

25. M.R. Foreman, S.S. Sherif, P. Török, Photon statistics in single molecule orientational imaging. Opt. Express **15**, 13597–13606 (2007)
26. B.R. Frieden, Maximum information data processing: application to optical signals. J. Opt. Soc. Am. **71**, 294–303 (1981)
27. B.R. Frieden, *Physics from Fisher Information: A Unification* (Cambridge University Press, Cambridge, 1998)
28. D. Gabor, Theory of communication. J. IEE **93**, 429–457 (1946)
29. T. Gehrels (ed.), *Planets Stars and Nebulae Studied with Photopolarimetry* (University of Arizona Press, Tuscon, 1974)
30. J.W. Goodman, *Introduction to Fourier Optics*, 2nd edn. (McGraw Hill, New York, 1996)
31. U. Gopinathan, T.J. Naughton, J.T. Sheridan, Polarization encoding and multiplexing of two-dimensional signals: application to image encryption. Appl. Opt. **45**, 5693–5700 (2006)
32. F. Goudail, Optimization of the contrast in active Stokes images. Opt. Lett. **34**, 121–124 (2009)
33. F. Goudail, A. Bénière, Optimization of the contrast in polarimetric scalar images. Opt. Lett. **34**, 1471–1473 (2009)
34. R.V.L. Hartley, Transmission of information. Bell Syst. Tech. J. **7**, 535–563 (1928)
35. Z. Hradil, Quantum-state estimation. Phys. Rev. A **55**, R1561–R1564 (1997)
36. Z. Hradil, J. Řeháček, Quantum measurement and information. Fortschr. Phys. **51**, 150–156 (2003)
37. International Organization for Standardization, and European Computer Manufacturers Association (1998) Optical interfaces for multichannel systems with optical amplifiers. ITU-T Recommendation G. 692, International Telecommunication Union
38. S.M. Kay, *Fundamentals of Statistical Signal Processing: Estimation Theory* (Prentice-Hall, Inc., London, 1993)
39. G.E. Keiser, A review of WDM technology and applications. Opt. Fiber Technol. **5**(1), 3–39 (1999)
40. W.D. Koek, N. Bhattacharya, J.J.M. Braat, V.S.S. Chan, J. Westerweel, Holographic simultaneous readout polarization multiplexing based on photoinduced anisotropy in bacteriorhodopsin. Opt. Lett. **29**, 101–103 (2004)
41. K. Küpfmüller, Uber einschwingvorgange in Wellen filtern. Elek. Nachrichtentech. **1**, 141–152 (1924)
42. D. Lara, C. Paterson, Stokes polarimeter optimization in the presence of shot and Gaussian noise. Opt. Express **17**, 21240–21249 (2009)
43. B. Laude-Boulesteix, A. De Martino, B. Drévillon, L. Schwartz, Mueller polarimetric imaging system with liquid crystals. Appl. Opt. **43**, 2824–2832 (2004)
44. A.W. Lohmann, R.G. Dorsch, D. Mendlovic, Z. Zalevsky, C. Ferreira, Space-bandwidth product of optical signals and systems. J. Opt. Soc. Am. A **13**, 470–473 (1996)
45. S.Y. Lu, R.A. Chipman, Interpretation of Mueller matrices based on polar decomposition. J. Opt. Soc. Am. A **13**, 1106–1113 (1996)
46. W. Lukosz, Optical systems with resolving powers exceeding the classical limit I. J. Opt. Soc. Am. **56**, 1463–1472 (1966)
47. Macías Romero, C. High numerical aperture Mueller matrix polarimetry and applications to multiplexed optical data storage. Ph.D. thesis, 2010
48. R. Mehra, Optimal input signals for parameter estimation in dynamic systems-survey and new results. IEEE Trans. Autom. Contr. **19**, 753–768 (1974)
49. D. Mendlovic, A.W. Lohmann, Spacebandwidth product adaptation and its application to superresolution: fundamentals. J. Opt. Soc. Am. A **14**, 2488–2493 (1997)
50. D.A.B. Miller, Communicating with waves between volumes: evaluating orthogonal spatial channels and limits on coupling strengths. Appl. Opt. **39**, 1681–1699 (2000)
51. E.H. Moore, On the reciprocal of the general algebraic matrix. Bull. Am. Math. Soc. **26**, 394–395 (1920)
52. J. Morio, F. Goudail, Influence of the order of diattenuator, retarder, and polarizer in polar decomposition of Mueller matrices. Opt. Lett. **29**, 2234–2236 (2004)

References

53. P.R.T. Munro, P. Török, Properties of high-numerical-aperture Mueller-matrix polarimeters. Opt. Lett. **33**, 2428–2430 (2008)
54. S.M. Nee, Error analysis for Mueller matrix measurement. J. Opt. Soc. Am. A **20**, 1651–1657 (2003)
55. M.A. Neifeld, Information, resolution, and space bandwidth product. Opt. Lett. **18**, 1477–1479 (1998)
56. H. Nyquist, Certain factors affecting telegraph speed. Bell Syst. Tech. J. **3**, 324–352 (1924)
57. H. Nyquist, Certain topics in telegraph transmission theory. AIEE Trans. **47**, 617–644 (1928)
58. R.J. Ober, S. Ram, E.S. Ward, Localization accuracy in single-molecule microscopy. Biophys. J. **86**(2), 1185–1200 (2004)
59. R. Oldenbourg, A new view on polarization microscopy. Nature **381**, 811–812 (1996)
60. R. Ossikovski, Analysis of depolarizing Mueller matrices through a symmetric decomposition. J. Opt. Soc. Am. A **26**, 1109–1118 (2009)
61. R. Penrose, A generalized inverse for matrices. Proc. Camb. Philos. Soc. **51**, 406–413 (1955)
62. R. Piestun, D.A.B. Miller, Electromagnetic degrees of freedom of an optical system. J. Opt. Soc. Am. A **17**, 892–902 (2000)
63. O. Popescu, C. Rose, D.C. Popescu, Maximising the determinant for a special class of block partitioned matrices. Math. Probl. Eng. **1**, 49–61 (2004)
64. S. Ram, E.S. Ward, R.J. Ober, Beyond Rayleigh's criterion: a resolution measure with application to single-molecule microscopy. Proc. Natl Acad. Sci. U S A **103**, 4457–4462 (2006)
65. A.C.S. Readhead, S.T. Myers, T.J. Pearson, J.L. Sievers, B.S. Mason, C.R. Contaldi, J.R. Bond, R. Bustos et al., Polarization observations with the cosmic background imager. Science **306**, 836–844 (2004)
66. T.J. Rothenberg, *Efficient Estimation with a priori Information* (Yale University Press, New Haven, 1973)
67. B.E.A. Saleh, M.C. Teich, Can the channel capacity of a light wave communication system be increased by the use of photon number squeezed light. Phys. Rev. Lett. **58**, 2656–2659 (1987)
68. B.E.A. Saleh, M.C. Teich, Information transmission with photon-number squeezed light. Proc. IEEE **80**(3), 451–460 (1992)
69. Savenkov, S. Eigenview on Mueller matrix models of homogeneous anisotropic media. In Advanced Polarimetric Instrumentation, Workshop 2009 (December 2009).
70. S.N. Savenkov, Optimization and structuring of the instrument matrix for polarimetric measurements. Opt. Eng. **41**, 965–972 (2002)
71. S.N. Savenkov, K.E. Yushtin, Mueller matrix elements error distribution for polarimetric measurements. Proc. SPIE **5158**, 251–259 (2003)
72. C.E. Shannon, A mathematical theory of communication. Bell Syst. Tech. J. 27, 379–423, 623–656 (1948)
73. A. Shields, Quantum optics: quantum logic with light, glass, and mirrors. Science **297**, 1821–1822 (2002)
74. B. Sick, B. Hecht, L. Novotny, Orientational imaging of single molecules by annular illumination. Phys. Rev. Lett. **85**, 4482–4485 (2000)
75. D. Slepian, Some comments on Fourier analysis, uncertainty and modeling. SIAM Rev. **25**, 379–393 (2000)
76. M. Smith, Optimization of a dual-rotating-retarder Mueller matrix polarimeter. Appl. Opt. **41**, 2488–2493 (2002)
77. L. Szilard, Über die entropieverminderun in einem thermodynamischen system bei eingriffen intelligenter wesen. Z. Phys. **53**, 593–604 (1929)
78. J. Tinbergen, Interstellar polarization in the immediate solar neighbourhood. Astron. Astrophys. **105**, 53–64 (1982)
79. P. Török, M. Salt, E.E. Kriezis, P.R.T. Munro, H.P. Herzig, C. Rockstuhl, Optical disk and reader therefor. Worldwide Patent WO2006/010882, 2006
80. N. Treps, V. Delaubert, A. Maître, J.M. Courty, C. Fabre, Quantum noise in multipixel image processing. Phys. Rev. A **71**, 013820 (2005)

81. J.S. Tyo, Optimum linear combination strategy for an n-channel polarization sensitive imaging or vision system. J. Opt. Soc. Am. A **15**, 359–366 (1998)
82. J.S. Tyo, Noise equalisation in Stokes parameter images obtained by use of variable-retardance polarimeters. Opt. Lett. **25**, 1198–1200 (2000)
83. J.S. Tyo, Design of optimal polarimeters: maximization of signal-to-noise ratio and minimization of systematic error. Appl. Opt. **41**, 619–630 (2002)
84. A. van der Sluis, Condition numbers and equilibration of matrices. Numer. Math. **14**, 14–23 (1969)
85. A.D. Whalen, *Detection of Signals in Noise* (Academic Press Inc., New York, 1971)
86. J. Zallat, S. Aïnouz, M.P. Stoll, Optimal configurations for imaging polarimeters: impact of image noise and systematic errors. J. Opt. A: Pure Appl. Opt. **8**, 807–814 (2006)

Chapter 6
Information in Polarisation Imaging

Observations always involve theory.
Edwin Hubble

6.1 Introduction

Information in both the natural and man-made world is frequently not spatially confined to a single point. Whilst, for example, studying the autofluorescence from a single molecule in a cell provides information with regards to that molecule, nothing is learnt about the processes and structure in the whole cell. To do so requires information to be collected from multiple locations. Such is the reason for the prevalence and success of imaging systems. In an optical context, a CCD can be used to record the intensity incident upon each pixel for instance. If located in the image plane of an optical microscope or telescope, information with regards to the object can then be extracted from the intensity readings.

Incorporating polarisation sensitive elements into existing optical imaging systems however affords the possibilities of polarisation imaging, in which either the polarisation state of the incident light is mapped (Stokes imaging) or the polarisation properties of the object are considered (Mueller imaging). For example the Pol-Scope developed by Oldenbourg et al. [39–41] is capable of measuring optical anisotropies in a sample, such as birefringence and diattenuation as may be of interest in molecular and crystallographic studies.

Great efforts are often spent on obtaining high quality images, which possess a high fidelity with the original object, so that a human observer can immediately interpret the image as desired, e.g. in medical diagnosis a clinician may look for cancerous regions of tissue. Such efforts often concentrate on improving the spatial resolution of the imaging system beyond the diffraction limit, often referred to as superresolution, so as to maximise the spatial information available. Tailoring the

M. R. Foreman, *Informational Limits in Optical Polarimetry and Vectorial Imaging*,
Springer Theses, DOI: 10.1007/978-3-642-28528-8_6,
© Springer-Verlag Berlin Heidelberg 2012

illumination field in the object plane is a popular technique used to this end, although numerical optimisation is normally used to determine the appropriate mask or field distribution to use. In Sect. 6.2 however a new analytic method, developed in part by the author, based upon an eigenfunction expansion of the Debye-Wolf diffraction integral is presented. In principle this method allows an arbitrary bandlimited field distribution to be specified in the focal region of a high numerical aperture lens and the appropriate pupil plane distribution to be calculated. Various additional considerations do however constrain the inversion to ensure physicality and practicality of the results including field specification, energy concentration and noise amplification. These topics are discussed fully in Sect. 6.3.1. Synthesis of arbitrary field distributions is of importance in a number of further applications including lithography, optical data storage, and atomic manipulation. Section 6.3.2 therefore considers a variety of examples, beyond that of superresolution, in which the inversion formalism is utilised.

High image fidelity in high NA imaging systems in general requires the component-wise point spread function to approximate a Dirac delta function. Even in the simple scenario of imaging a dipole this condition is however not fulfilled, due to the mixing of field components upon focusing (see Sect. 4.5). Information theory fortunately provides an alternative strategy for assessing images, whereby the final image is treated as a message from which information about the object can be extracted [13]. The latter half of this chapter (Sect. 6.4) is therefore dedicated to the use of the Fisher information concepts developed thus far, to characterise the performance of imaging systems. Although initially, the increased informational capabilities of imaging systems will be formally proven, particular attention will be given to the imaging properties of a simple polarisation microscope imaging dipole sources. This example is adopted primarily for two reasons; firstly, dipole sources constitute the elementary object from which more complex optical systems can be modelled [61]; whilst secondly, polarisation microscopes play a key role in the readout of the polarisation multiplexed optical data storage system presented in Chap. 7. Insight can thus be gained into these systems.

Finally the chapter closes with Sect. 6.5, in which potential accuracy gains achievable when incorporating a priori information into image processing routines, is briefly investigated. As contrasted to the probabilistic a priori knowledge considered in Chap. 5, physical constraints, as dictated by Maxwell's equations, are employed. Significant improvements will be seen to be possible.

6.2 Eigenfunction Expansion of the Debye-Wolf Diffraction Integral

As discussed in Sect. 4.5 the Debye-Wolf diffraction integral is routinely used to describe the focusing of light in high numerical aperture systems. With a view to simplifying the Debye-Wolf integral the formulae of Watson [62], Gradshteyn and Ryzhik [19] and Agrawal and Pattanayak [1], were used by Török et al. to express

6.2 Eigenfunction Expansion of the Debye-Wolf Diffraction Integral

both the in-focus and defocus terms by means of a series of analytic functions [58]. Kant also reported a series expansion of the diffraction integrals using Gegenbauer polynomials [26] and Sherif and Török [52] further reported an eigenfunction representation of the so-called I integrals of Richards and Wolf [45, 66]. In reality however these expansions do not reveal anything about the physical nature of the problem, but instead merely provide a simplified means to calculate the diffraction integrals when computational time is deemed to be of significance.

Braat et al. obtained the field components in the focal region as a series using Nijboer-Zernike functions [5], whilst Sheppard and Török obtained the field components as a multipole expansion [48]. These two expansions are more physical than those listed above because the Nijboer-Zernike expansion aims at obtaining formulae where the incident and focused fields are represented in terms of aberration functions. This representation has immediate significance in the study of realistic focusing systems with aberrations present. Nevertheless, the multipole expansion may be regarded as the most physical representation because the focused field is represented in terms of physically realisable multipoles.

In this section, a new expansion of the electric field components in the focal region of a high NA lens, developed by the author in collaboration with Sherif et al. [50], is presented. This expansion is in terms of Bessel functions and the generalised prolate spheroidal functions which are eigenfunctions of the two-dimensional finite Hankel transform. Whilst Bessel functions are likely to be familiar to the reader, the generalised prolate spheroidal functions are less commonly encountered and hence an introduction is given in Appendix B. As is discussed in Sect. 6.2.2, the presented eigenfunction expansion has many optimal, desirable and physical properties, including maximum energy packing properties, bandlimited basis functions, separability in cylindrical coordinates, and fast convergence in the azimuthal, radial and axial directions. Physically the dominant modes are furthermore closely related to the resolution of the optical system. Consequently, it will be seen that the eigenfunction expansion provides a simple and natural way to carry out both forward and inverse analysis of high NA focusing systems. In contrast to earlier work [52] which allowed only for 1D apodisation techniques [51], the current expansion can be used to implement 2D apodisation and masking techniques to synthesise arbitrary fields in the focal region of a high NA focusing system, as discussed by the author in [14].

6.2.1 Derivation of the Eigenfunction Expansion

Recall from Sect. 4.5 that the Debye-Wolf integral can be written

$$\mathbf{E}(\rho, \varphi, z) = -\frac{if}{\lambda} \int_0^{2\pi} \int_0^{\alpha} \mathbf{e}(\theta, \phi) \exp\left[ik\rho \sin\theta \cos(\phi - \varphi)\right] e^{ikz\cos\theta} \sin\theta d\theta d\phi, \tag{6.1}$$

where (ρ, φ, z) defines a position in the focal region of a lens (in air) of focal length f and numerical aperture $\sin \alpha$, illuminated with light of wavelength $\lambda = 2\pi/k$. The coordinates (θ, ϕ) define the direction of a ray, with polarisation given by $\mathbf{e}(\theta, \phi)$ on the Gaussian reference sphere of the lens (see Fig. 4.2).

To begin the derivation, the defocus term, $\exp[ikz \cos \theta]$, in Eq. (6.1) is rewritten, using the Jacobi-Anger expansion [2], as

$$\exp[ikz \cos \theta] = \sum_{m=-\infty}^{\infty} i^m J_m(kz) \exp[im\theta] = \sum_{m=-\infty}^{\infty} J_m(kz) \exp\left[im\left(\frac{\pi}{2} - \theta\right)\right], \tag{6.2}$$

where $J_m(\cdots)$ is the Bessel function of the first kind of order m. Substituting Eq. (6.2) in Eq. (6.1), gives

$$\mathbf{E}(\rho, \varphi, z) = -\frac{if}{\lambda} \sum_{m=-\infty}^{\infty} J_m(kz) \int_0^{2\pi} \int_0^{\alpha} \mathbf{e}(\theta, \phi) \exp[ik\rho \sin \theta \cos(\phi - \varphi)]$$
$$\times \exp\left[im\left(\frac{\pi}{2} - \theta\right)\right] \sin \theta \, d\theta \, d\phi, \tag{6.3}$$

Equation (6.3) can be simplified through the coordinate transformation $u = \sin \theta$ whereby $\cos \theta = \sqrt{1 - u^2}$ and $d\theta = du/\cos \theta$. Substituting this transformation into Eq. (6.3) yields

$$\mathbf{E}(\rho, \varphi, z) = -\frac{if}{\lambda} \sum_{m=-\infty}^{\infty} J_m(kz) \int_0^{2\pi} \int_0^{u_\alpha} \mathbf{a}(u, \phi) \exp[ik\rho u \cos(\phi - \varphi)] u \, du \, d\phi, \tag{6.4}$$

where $u_\alpha = \sin \alpha$ and

$$\mathbf{a}(u, \phi) = \frac{\mathbf{e}(u, \phi)}{\sqrt{1 - u^2}} \exp\left[im\left(\frac{\pi}{2} - \sin^{-1} u\right)\right]. \tag{6.5}$$

The function $\mathbf{a}(u, \phi)$ is space-limited, so it can be expanded in terms of generalised circular prolate spheroidal functions (see Appendix B), which are eigenfunctions of the two-dimensional finite Hankel transform [23, 54], viz.

$$\mathbf{a}(u, \phi) = \sum_{N=-\infty}^{\infty} \sum_{n=0}^{\infty} \mathbf{A}_{m,N,n} \Phi_{|N|,n}(u, c) \exp(iN\phi), \tag{6.6}$$

where

$$\mathbf{A}_{m,N,n} = \frac{1}{2\pi \lambda_{|N|,n}} \int_0^{2\pi} \int_0^{u_\alpha} \mathbf{a}(u, \phi) \Phi_{|N|,n}(u, c) \exp(-iN\phi) u \, du \, d\phi, \tag{6.7}$$

6.2 Eigenfunction Expansion of the Debye-Wolf Diffraction Integral

are vector expansion coefficients and c is a parameter equal to, or larger than, the radial space-bandwidth product (see Appendix B) of $\mathbf{a}(u, \phi)$ for $0 < \phi \leq 2\pi$ [15].

Upon substitution of Eq. (6.6) into Eq. (6.4) and changing the order of the integration and summation,

$$\mathbf{E}(\rho, \varphi, z) = -\frac{if}{\lambda} \sum_{m=-\infty}^{\infty} J_m(kz) \sum_{N=-\infty}^{\infty} \sum_{n=0}^{\infty} \mathbf{A}_{m,N,n}$$

$$\times \int_0^{2\pi} \int_0^{u_\alpha} \Phi_{|N|,n}(u, c) \exp(iN\phi) \exp[ik\rho u \cos(\phi - \varphi)] \, u \, du \, d\phi$$

(6.8)

is obtained. Using the expansion $\exp[ik\rho u \cos \phi] = \sum_{Q=-\infty}^{\infty} i^Q J_Q(k\rho u)$ $\exp(iQ\phi)$ and the fact that $\cos(\phi - \varphi) = \cos(\varphi - \phi)$ and $\int_0^{2\pi} \exp(i(N-Q)\phi) \, d\phi = 2\pi \delta_{NQ}$ produces

$$\mathbf{E}(\rho, \varphi, z) = -ikf \sum_{m=-\infty}^{\infty} J_m(kz) \sum_{N=-\infty}^{\infty} i^N \exp(iN\varphi)$$

$$\times \sum_{n=0}^{\infty} \mathbf{A}_{m,N,n} \int_0^{u_\alpha} \Phi_{|N|,n}(u, c) J_N(k\rho u) \, u \, du.$$

(6.9)

Further noting that for $N < 0$ the relation $J_N(x) = (-1)^N J_{|N|}(x)$ holds and that $i^N (-1)^N = i^{-N} = i^{|N|}$ yields

$$\mathbf{E}(\rho, \varphi, z) = -ikf \sum_{m=-\infty}^{\infty} J_m(kz) \sum_{N=-\infty}^{\infty} i^{|N|} \exp(iN\varphi)$$

$$\times \sum_{n=0}^{\infty} \mathbf{A}_{m,N,n} \int_0^{u_\alpha} \Phi_{|N|,n}(u, c) J_{|N|}(k\rho u) \, u \, du.$$

(6.10)

Using the eigen-equation [15]

$$\int_0^{r_0} \Phi_{|N|,n}(c, r) J_{|N|}(wr) \, r \, dr = (-1)^n \left(\frac{r_0}{\Omega}\right) \sqrt{\lambda_{|N|,n}} \Phi_{|N|,n}\left(c, \frac{r_0 w}{\Omega}\right),$$

(6.11)

where $\Omega = \omega_{\max}$, with the appropriate replacements $r_0 = u_\alpha$, $\omega = k\rho$ and $\Omega = k\rho_{\max}$, where $\rho_{\max}$ defines the transverse field of view in the focal region (see Appendix B), finally yields the expansion

$$\mathbf{E}(\rho, \varphi, z) = -ikf\left(\frac{u_\alpha}{k\rho_{\max}}\right)$$
$$\times \sum_{m=-\infty}^{\infty} \sum_{N=-\infty}^{\infty} \sum_{n=0}^{\infty} i^{|N|} \mathbf{A}_{m,N,n} (-1)^n \sqrt{\lambda_{|N|,n}} \, \Phi_{|N|,n}\left(c, \frac{u_\alpha \rho}{\rho_{\max}}\right) J_m(kz) \, e^{iN\varphi}, \tag{6.12}$$

where $c = u_\alpha k \rho_{\max}$.

In the preceding derivation no assumption was made as to the form of the incident polarisation $\mathbf{e}(u, \phi)$. Consequently the formulation given is suitable for representing the field in the focal region produced by an arbitrary illumination. For example the field on the reference sphere $\mathbf{e}(u, \phi)$ could be given in the form (c.f. Eq. (4.47))

$$\mathbf{e}(u, \phi) = g(u, \phi) e^{i\Psi(u,\phi)} \mathbb{Q}(u, \phi) \widetilde{\mathbf{E}}(u, \phi), \tag{6.13}$$

where $g(u, \phi)$ and $\Psi(u, \phi)$ describe an amplitude and phase variation that could be introduced to the incident field distribution by pupil plane masks and $\mathbb{Q}(u, \phi)$ describes the action of the lens and maps the field to the Gaussian reference sphere as discussed in Sect. 4.5.

6.2.2 Properties of the Eigenfunction Expansion

On examining the eigenfunction expansion of the Debye-Wolf integral (Eq. (6.12)), a number of desirable, optimal and physical properties can be identified. Firstly, the component functions $J_m(kz)$, $\exp(iN\varphi)$, $\Phi_{|N|,n}(c, u_\alpha \rho/\rho_{\max})$ are separable in cylindrical coordinates which could simplify analysis involving fields in the focal region of a high NA focusing system.

Secondly, the double integral in Eq. (6.8) represents a finite Hankel transform of the generalised prolate spheroidal functions. As a linear operator, this transform takes its simplest possible form, i.e., diagonal, through its eigen representation given in Eq. (6.12).

Thirdly, one of the defining properties of the generalised prolate spheroidal functions is that $\Phi_{|N|,0}(r, c)$ maximise the fractional energy within a circular region of radius r_0 over the class of all bandlimited functions [15]. Thus the summation over N has maximum energy packing properties in both the radial and azimuthal direction when $n = 0$. More generally, the eigenvalues, $\lambda_{|N|,n}$, describe the fractional energy within the circular region for each generalised prolate spheroidal function.

Fourthly, as shown in Fig. 6.1a, the eigenvalues of the generalised prolate spheroidal functions are seen to decrease monotonically to very small values, compared to their initial values, after certain orders $|N| \geq |N_0|$ and $n \geq n_0$. Furthermore, as shown in Fig. 6.1b, Bessel functions of the same argument, but of increasing orders, also decrease to very small values after some order $m \geq m_0$. Computationally

6.2 Eigenfunction Expansion of the Debye-Wolf Diffraction Integral

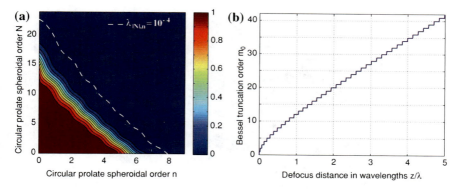

Fig. 6.1 a Monotonically decreasing eigenvalues of the circular prolate spheroidal functions as order (N, n) increases $(c = 20)$. A value of $\lambda_{|N|,n} = 10^{-4}$ was used to determine a suitable truncation point for the infinite series inherent in the eigenfunction expansion; **b** finite summation limit for Bessel terms required for different defocus distances again based on a cutoff point of 10^{-4}

these properties are desirable since it implies fast convergence of the eigenfunction expansions in the azimuthal, radial and axial directions.

The presented representation of the field in the focal region of a focusing system uses scalar basis functions and vector coefficients. Equivalently this can be viewed as a separate expansion for each field component, for which the basis field distributions are unaltered by the focusing operation. In Fig. 6.2 the in-focus ($z = 0$) eigenfunctions, which reduce to the generalised prolate spheroidal functions are plotted. Defocused eigenfunctions however are further modulated by a Bessel function dependent on the axial coordinate. For low numerical aperture systems the polarisation properties of light become less important, often allowing a scalar treatment to be used. Under such circumstances the field distributions shown in Fig. 6.2 can be interpreted as the true eigenfunctions of the focusing operation. At higher numerical apertures the distributions shown are not strictly eigenfunctions of the Debye-Wolf integral since they are scalar functions, however they do remain eigenfunctions on a component-wise basis, i.e. if x-polarised light with amplitude distribution given by the (N, n)th order were focused, the x component of the output field would also have the same (albeit scaled) distribution.

Higher order functions are seen to contain a larger fraction of energy in the sidelobe structure, which itself becomes progressively more complicated as the order increases (more nodal points). Furthermore, only the $N = 0$ modes are seen to possess a central focal spot. Put another way, the higher order functions contain higher spatial frequencies however energy is not concentrated as efficiently into the central region. Consequently more complicated masking optics will, in general, require more terms to be calculated in the eigenfunction expansion to accurately determine the field in the focal region.

124   6 Information in Polarisation Imaging

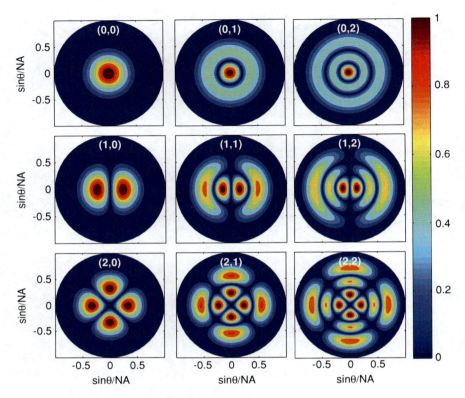

Fig. 6.2 Absolute magnitude of the generalised prolate spheroidal functions of order (N, n). Distributions represent 2D component-wise eigenfunctions for the operator describing the transformation of a field distribution in the back focal plane of a high NA focusing lens to the focal plane ($z = 0$), such that the summation over m in Eq. (6.12) can be safely neglected

Finally in Fig. 6.3 the variation of the eigenvalues, $\lambda_{|N|,n}$, with NA of the focusing system is shown. The decrease of the eigenvalue for a given order (N, n) as the NA decreases is clearly evident. This dependence means that higher orders are energetically less significant in the focused distribution and hence the dominant orders are those with lower spatial frequencies. The dominant modes in the focal region hence provide a characterisation of the resolution of the optical system.

6.2.3 Numerical Examples

To verify the validity of the eigenfunction expansion, Eq. (6.12) is used to calculate the field distributions in the Gaussian focal plane and a defocused plane for a horizontally polarised incident beam, and compared to the corresponding

6.2 Eigenfunction Expansion of the Debye-Wolf Diffraction Integral

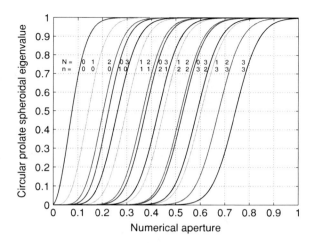

Fig. 6.3 Variation of eigenvalues $\lambda_{|N|,n}$ of the generalised prolate spheroidal functions as the numerical aperture of the focusing optical system is continuously changed. Different plots correspond to different prolate orders (N, n)

distributions obtained by the direct evaluation (numerical integration) of the Debye-Wolf integral (Fig. 4.3). Assuming the availability of tabulated values or computer routines[1] to evaluate Bessel functions and circular prolate spheroidal functions [12, 30], the main task to evaluate the eigenfunction expansion is to determine the space-bandwidth product c and suitable finite limits for the three summations in Eq. (6.12). The parameter c has to be equal to or larger than the radial space-bandwidth product of the function $\mathbf{a}(u, \phi)$ for all values $0 < \phi \leq 2\pi$. In addition $c = u_\alpha k \rho_{max}$, thus for a given NA, c determines the radial field of view, ρ_{max}, in the plane of interest. For the following numerical examples, a value of $c = 20$ was found to satisfy these two requirements.

From Fig. 6.1a truncation orders of $|N_0| = 23$ and $n_0 = 8$ are found for $c = 20$, whilst from Fig. 6.1b, and assuming a defocus distance of $z = \lambda$, $m_0 = 14$ is found to be an appropriate limit for the summation with respect to m.

Figures 6.4 and 6.5 show the pointwise relative error between the optical distributions due to a linearly polarised incident beam, at the Gaussian focal plane and at a defocused plane, $z = \lambda$, respectively, obtained by evaluating the eigenfunction expansion, Eq. (6.12), with finite summation limits and by direct integration. The actual calculated optical distributions are also shown in the insets.

To confirm that these optical distributions are indeed equal to the ones obtained by direct evaluation of the Debye-Wolf integral, an overall percentage error factor, Δ, is defined as

$$\Delta = \left(\sum_k \sum_l \left| I_{k,l}^{direct} - I_{k,l}^{expansion} \right| \bigg/ \sum_k \sum_l I_{k,l}^{direct} \right) \cdot 100, \qquad (6.14)$$

---

[1] Paul Abbott and Peter Falloon from the Physics Department of the University of Western Australia kindly provided the *Mathematica* code necessary to compute Slepian's generalised spheroidal functions, for which the author is particularly grateful.

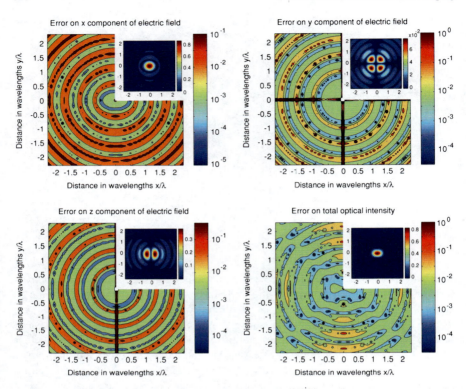

Fig. 6.4 Pointwise relative error (defined as $|X_{k,l}^{\text{direct}} - X_{k,l}^{\text{expansion}}|/|X_{k,l}^{\text{direct}}|$) of the electric field components ($X_{k,l} = E_j(\rho_{k,l})$) and intensity ($X_{k,l} = I(\rho_{k,l})$) distributions at the Gaussian focal plane, when calculated using the eigenfunction expansion, as compared to direct integration (*x* polarised incident illumination, NA = 0.966). *Strong horizontal* and *vertical lines* are seen due to zeros in the field distributions. *Insets* show absolute magnitude of the field and intensity distributions calculated using the eigenfunction representation

where $I_{k,l}^{\text{direct}}$ and $I_{k,l}^{\text{expansion}}$ are the intensities at point (k, l) obtained by evaluating the Debye-Wolf integral and the eigenfunction expansion respectively. On applying Eq. (6.14) to the above numerical examples, the associated errors Δ are found to be 0.0253 and 0.065% at the Gaussian focal plane (Fig. 6.4) and at a defocus distance $z = \lambda$ (Fig. 6.5), respectively. This overall error is very small, bearing in mind the relatively small number of terms used to evaluate Eq. (6.12). Furthermore it is found the variation of the total percentage error as the NA is increased from 0 to 0.966 is of order 0.003% and is hence negligible.

6.3 Inversion of the Debye-Wolf Diffraction Integral

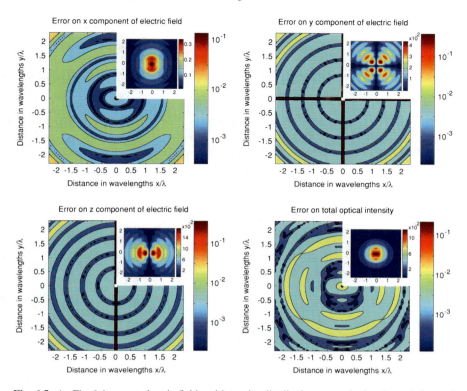

Fig. 6.5 As Fig. 6.4 except electric field and intensity distributions are calculated at a defocused plane ($z = \lambda$)

6.3 Inversion of the Debye-Wolf Diffraction Integral

Synthesis of arbitrary field distributions in optical systems is useful for a wide variety of applications including lithography [11], optical data storage [6], atomic manipulation [46] and polarisation microscopy [25]. A significant number of alternative methods by which to produce a desired field distribution exist, such as apodisation or phase masks [9], polarisation structuring [8] and computer generated holograms [17, 31]. Numerical optimisation is however normally used to determine the appropriate mask or field distribution to use [68]. To the best of the author's knowledge there does not currently exist an analytic method to invert the Debye-Wolf integral in the literature. The eigenfunction expansion presented above however allows this situation to be remedied.

In principle the new method allows an arbitrary field distribution to be specified in the focal region of a high NA lens and the appropriate weighting function, or equivalently the pupil plane distribution, to be calculated. Furthermore, due to the simple form of the inversion, it is still amenable to numerical optimisation should extra constraints need to be introduced to the system, allowing existing optimisation tools to be exploited. Such constraints may include the pixelation of masking optics, a feature often encountered when using spatial light modulators (SLMs) [36] for example, which limits the level of fine structure producible in any physical mask and thus any pupil plane field distribution.

Given Eq. (6.12) for the field in the focal region of a high NA lens and the orthogonality of the generalised prolate spheroidal functions as described by Eq. (B.4) it is possible to invert the Debye-Wolf integral as follows. Consider multiplying both sides of Eq. (6.12) by the conjugate of the generalised prolate spheroidal function of order (Q, q) and integrating over the field of view (as set by ρ_{max}) in the focal region.[2] This yields the result

$$\int_0^{2\pi} \int_0^{\rho_{max}} \mathbf{E}(\rho, \varphi, z) \, \Phi_{|Q|,q}\left(\frac{u_\alpha \rho}{\rho_{max}}\right) \exp(-iQ\varphi) \rho \, d\rho \, d\varphi$$
$$= -2\pi i k f \left(\frac{k\rho_{max}}{u_\alpha}\right) \sum_{m=-\infty}^{\infty} \sum_{N=-\infty}^{\infty} \sum_{n=0}^{\infty} i^{|N|} \mathbf{A}_{m,N,n}(-1)^n \lambda_{|N|,n}^{3/2} \delta_{QN} \delta_{qn} J_m(kz). \tag{6.15}$$

The Kronecker deltas eliminate all but a single term within the double summation over N and n, namely the term for which $Q = N$ and $q = n$. Thus

$$\mathbf{B}_{N,n} = -ikf\left(\frac{k\rho_{max}}{u_\alpha}\right) i^{|N|}(-1)^n \lambda_{|N|,n}^{1/2} \sum_{m=-\infty}^{\infty} \mathbf{A}_{m,N,n} J_m(kz), \tag{6.16}$$

where

$$\mathbf{B}_{N,n} = \frac{1}{2\pi \lambda_{|N|,n}} \int_0^{2\pi} \int_0^{\rho_{max}} \mathbf{E}(\rho, \phi, z) \, \Phi_{|N|,n}\left(\frac{u_\alpha \rho}{\rho_{max}}\right) \exp(-iN\varphi) \rho \, d\rho \, d\varphi. \tag{6.17}$$

Trivial algebraic rearrangement of Eq. (6.16) yields an infinite set of linear equations,

$$\sum_{m=-\infty}^{\infty} J_m(kz) \mathbf{A}_{m,N,n} = \frac{i}{kf}\left(\frac{u_\alpha}{k\rho_{max}}\right) \frac{(-1)^n}{i^{|N|}} \lambda_{|N|,n}^{-1/2} \mathbf{B}_{N,n}, \tag{6.18}$$

which unfortunately cannot be solved analytically to determine the desired coefficients $\mathbf{A}_{m,N,n}$, but can however form the basis for numerical optimisation techniques, an example of which is given in Sect. 6.3.2. Unique solution can however be achieved on the focal plane, i.e. when $z = 0$, whereby

---

[2] Mathematically the same inversion procedure can be followed using integration over an infinite plane in the focal region, however here the mathematics is demonstrated using a restricted domain since such integrations are more suitable for numerical routines.

6.3 Inversion of the Debye-Wolf Diffraction Integral

$$J_m(kz)\Big|_{z=0} = \begin{cases} 1 & \text{for } m = 0 \\ 0 & \text{otherwise} \end{cases}, \quad (6.19)$$

yielding the simple relation

$$\mathbf{A}_{N,n} = \frac{i}{kf}\left(\frac{u_\alpha}{k\rho_{\max}}\right)\frac{(-1)^n}{i^{|N|}}\lambda_{|N|,n}^{-1/2}\mathbf{B}_{N,n}. \quad (6.20)$$

For in-focus distributions the expansion of the defocus term (Eq. 6.2) is not required in the derivation of Eq. (6.12) and hence the subscript m has now been dropped. This equation shows that the coefficients of the expansion of the weighting function are merely a scaled version of the coefficients of the expansion of the field in the focal plane as would be expected for an eigenfunction expansion. This is the basic inversion formula for the Debye-Wolf integral.

6.3.1 Some Notes on Inversion

Although a formula to invert the Debye-Wolf integral has now been derived, numerous problems may be encountered if it is used incorrectly. In this section some principles and caveats to use of Eq. (6.20) are thus presented.

6.3.1.1 Degrees of Freedom

The underlying purpose of inversion of the Debye-Wolf integral is to provide a means by which to generate a desired field distribution. As it stands Eq. (6.20) describes how to find the expansion coefficients of all three components of $\mathbf{a}(u, \phi)$, if all field components were specified in the focal region. A naïve approach such as this would however not guarantee physicality or realisability. Maxwell's equations mean that at best only two field components can be specified in the focal region and used for inversion, however there is no restriction on which components are chosen.

Furthermore, since some form of additional optics, e.g. a pupil plane mask, must be introduced into the system so as to modify the weighting function, there are additional constraints on the specification of the electric field on the focal plane. These constraints arise from the degrees of freedom inherent in the optics introduced. To illustrate this point consider use of an apodisation mask in the exit pupil of the system. This introduces only a single degree of freedom to the system, that is to say, only the amplitude of the field in the pupil plane can be modified and not its phase. In turn, this translates to the requirement that the field component specified in the focal region must be complex Hermitian, such that $E_j(\rho) = E_j^*(-\rho)$. Combination of a phase and apodisation mask would however provide two degrees of freedom, allowing an arbitrary phase and amplitude profile to be specified for a single field component in the focal plane. A polarisation mask similarly provides two degrees of freedom hence two field components in a focal plane can be specified without constraints if amplitude, phase and polarisation masks are used concurrently. Assuming more degrees of freedom than are present in a particular optical setup will lead to inconsistent inversion results that will not reproduce the desired field distribution and should hence be avoided.

6.3.1.2 Field Specification Away from the Focal Plane

Inversion was previously restricted to the focal plane since it is not possible to solve a set of $(2N_0+1) \times (n_0+1)$ equations for $(2m_0+1) \times (2N_0+1) \times (n_0+1)$ unknowns uniquely. This restriction can however be circumvented under certain circumstances.

If only a single field component is specified on a plane in the focal region, but not necessarily the focal plane it is then possible to propagate this field to the focal plane by means of scalar techniques such as the angular spectrum method. Once the field on the focal plane has been obtained in this manner Eq. (6.20) can be used as prescribed above.

Alternatively, if two field components are specified then it is again possible to propagate the field to the focal plane, however vector formulations, such as e- and m-theory must instead be used [27].

6.3.1.3 Extrapolation and Encircled Energy

Specification of a desired field distribution in the focal plane over an infinite region is not only impractical, but also superfluous to physical requirements. As such, the inversion formula assumes the field is specified over a finite region of maximum extent $\rho_{\max}$. So as to ensure the completeness of the prolate functions over the specification area it is necessary to use the appropriate space-bandwidth product $c = ku_\alpha \rho_{\max}$ when calculating the coefficients from Eq. (6.17).

Perhaps the most important issue arising from only specifying a finite area is the resulting behaviour of the field outside of this region. Superresolution is a concept in which the synthesis of a focal spot smaller than the Rayleigh diffraction limit is attempted [42, 51], and provides a good example to highlight how this can be of relevance.

Consider specifying a sub-diffraction focal spot in E_x as shown in Fig. 6.6a over a circle of radius ~ 1.6 times that of the Airy disc. Expansion of the specified field in terms of generalised prolate spheroidal functions as per Eqs. (6.12) and (6.17) allows extrapolation of the field beyond this region since [15]

$$f(\rho) = \sum_{n=0}^{\infty} \lambda_{N,n}^{-1} \Phi_{N,n}(\rho) \int_0^{\rho_{\max}} f(\rho')\Phi_{N,n}(\rho')\rho'd\rho' \quad \text{for} \quad \begin{array}{l} \rho > 0 \text{ if } N > 0 \\ \rho \geq 0 \text{ if } N = 0 \end{array}.$$
(6.21)

This equation states that with knowledge of the function $f(\rho)$ over a finite region $0 \leq \rho \leq \rho_0$ it is possible to extrapolate to all values of $\rho > 0$ and is a consequence

6.3 Inversion of the Debye-Wolf Diffraction Integral 131

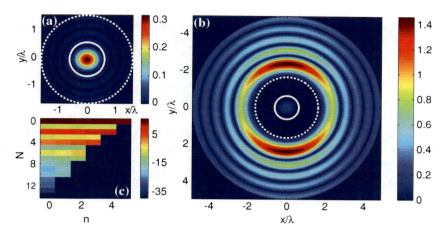

Fig. 6.6 When trying to produce a structure smaller than the diffraction limit as shown in (**a**); significant energy is pushed outside the specification area (**b**); as bounded by the *dashed line*. The *solid white line* in (**a**) and (**c**) shows the size of the Airy disc. This can be understood from the high order contributions to the specified field as shown in (**c**) which plots $\ln(|A_{N,n}|^2 \lambda_{N,n})$ to allow comparison between all modes

of the duality of completeness and orthogonality of the circular prolate spheroidal functions (see Appendix B).

The resultant field from extrapolation of the field distribution of Fig. 6.6a is shown in Fig. 6.6b. It can be seen that a significant fraction of energy is pushed out into the sidelobe/peripheral structures. When calculated, the encircled energy is found to be 6.86×10^{-4}. This behaviour arises since the high order modes contribute significantly as shown in Fig. 6.6c and thus energy is pushed out of the specification area as discussed in Sect. B.1.4.

6.3.1.4 Noise Amplification

Hadamard defined a number of criteria which a mathematical problem must meet to be well-posed [69], namely that a unique solution exists that depends continuously on the data, i.e. is stable. Inverse problems, such as that considered here, are however in general ill-posed, i.e. violate one or more of these conditions. Predominantly such a situation arises due to sensitivity to the initial data, which in the problem under consideration is a specified field distribution. In terms of inversion of the Debye-Wolf integral, errors in the specified field arise since the series expansion must be truncated for computational purposes.

Again the concept of the condition number of an inversion problem, κ, can be used to quantity the amplification of noise and errors in the initial data to the final inversion [21]. When using an eigenfunction inversion method the condition number can be defined as the ratio of the largest and smallest non-zero eigenvalue,[3] that is

$$\kappa = \frac{\lambda_{0,0}}{\lambda_{|N_0|,n_0}} \approx \frac{1}{\lambda_{|N_0|,n_0}}, \qquad (6.22)$$

where $\lambda_{|N_0|,n_0}$ is the value of the smallest eigenvalue used in the truncated series expansion. It is thus advisable to use orders that lie within or close to the plateau of eigenvalues of Fig. 6.1a to reduce noise amplification. Since small eigenvalues (high orders) correspond to high frequency components better inversion will be obtained for smoother, slower varying fields.

6.3.1.5 Pixelation

A final consideration that may arise in many practical systems is that of pixelation. Exact reproduction of the required pupil plane field distribution is generally not possible in practice due to the pixelated nature of the liquid crystal SLMs often used to implement complex masks [36] and as such an error on the focused field distribution is introduced. Choosing individual pixel values so as to minimise this error is then a further problem. Fortunately, since focusing is a unitary transformation i.e. one in which the inner product is conserved, minimisation of the root mean square (RMS) error in the focal plane is equivalent to minimising the RMS error in the exit pupil between the ideal and the pixelated mask. Doing so requires that the (j, k)th pixel of the SLM be set such that the output field is the average of the ideal profile over the domain Π_{jk} of the pixel i.e.

$$\widetilde{\mathbf{E}}_{jk} = \frac{1}{S_{jk}} \iint_{\Pi_{jk}} \widetilde{\mathbf{E}}(u, \phi) u du d\phi, \qquad (6.23)$$

where S_{jk} denotes the area of the (j, k)th pixel.

6.3.2 Examples

6.3.2.1 Superresolution

In this section, a few examples are given so as to illustrate the inversion procedure, the first of which again considers the topic of superresolution. The use of apodising pupil plane masks for such a purpose has previously been considered [9, 22], however use of a polarisation structured beam to obtain superresolution is attempted here.

---

[3] This definition of the condition number differs from that used in Chap. 5, however is more appropriate to an eigenfunction analysis.

6.3 Inversion of the Debye-Wolf Diffraction Integral

In an attempt to reduce the width of the intensity profile, consider specifying the E_x field component as a Dirac delta function centered on the origin. Only one component of the focused field is specified since this introduces two degrees of freedom into the inversion problem as required for polarisation structuring. E_x is hence written in the form

$$E_x(\rho, \varphi, 0) = \frac{1}{\rho} \delta\left(\frac{u_\alpha \rho}{\rho_{max}}\right) \delta(\varphi), \tag{6.24}$$

$$= \sum_{N=-\infty}^{\infty} \sum_{n=0}^{\infty} \lambda_{|N|,n}^{-1} \Phi_{|N|,n}\left(c, \frac{u_\alpha \rho}{\rho_{max}}\right) \Phi_{|N|,n}(c, 0) \exp(iN\varphi), \tag{6.25}$$

where the second step has used the completeness property of the generalised prolate spheroidal functions (Eq. B.4). Applying the inversion formula (Eq. 6.20) and noting $\Phi_{|N|,n}(0) = 0$ for $N \neq 0$, immediately gives

$$A^x_{N,n} = \begin{cases} \frac{i}{kf}\left(\frac{u_\alpha}{k\rho^{max}}\right)(-1)^n \lambda_{0,n}^{-1/2} \Phi_{0,n}(c, 0) & \text{for } N = 0 \\ 0 & \text{for } N \neq 0 \end{cases}, \tag{6.26}$$

where the superscript x denotes the x-component of $\mathbf{A}_{m,N,n}$.

Using Eq. (6.13) and noting that for a purely polarised structured beam of unit intensity $\widetilde{E}_x^2 = 1 - \widetilde{E}_y^2$, a quadratic equation in terms of $\widetilde{E}_x$ can be found, from which the required incident field distributions can be calculated. In practice however this method does not achieve superresolution for the simple reason that insufficient control is exerted on the y and z components of the focused field. As such when a delta function is specified for the x component, energy is pushed into the y component. The resultant focused distribution is then essentially identical to that of a uniformly y polarised beam for which there is no resolution improvement.

Consider then specifying both the E_x and E_y focused field components to be Dirac delta functions. By the same logic this means $A^x_{N,n} = A^y_{N,n}$ as given by Eq. (6.26). Since $a^j(u, \phi) = \sum_{n=0}^{\infty} A^j_{0,n} \Phi_{0,n}(c, u)$ (for $j = x, y, z$) the required incident field distributions can be found using the inverse of Eqs. (6.13) and (6.26) and are given by

$$\widetilde{E}_x(u, \phi) = \frac{1}{\sqrt{(1-u^2)}} \frac{Q_{21} - Q_{22}}{Q_{11}Q_{22} - Q_{12}Q_{21}} \sum_{n=0}^{\infty} A^x_{0,n} \Phi_{0,n}(c, u),$$

$$\widetilde{E}_y(u, \phi) = \frac{1}{\sqrt{(1-u^2)}} \frac{Q_{12} - Q_{11}}{Q_{11}Q_{22} - Q_{12}Q_{21}} \sum_{n=0}^{\infty} A^x_{0,n} \Phi_{0,n}(c, u), \tag{6.27}$$

where Q_{pq} denotes the (p, q)th element of $\mathbb{Q}$.

Having specified two field components on the focal region plane means there are four degrees of freedom within the system. Such a situation could correspond to the combination of polarisation structuring, apodisation and phase modulation in the

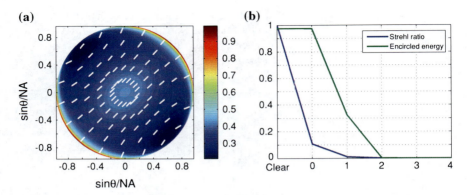

Fig. 6.7 **a** Colour plot showing the transmittance of the apodising mask in the pupil plane, whilst *white lines* represent the plane of oscillation of the electric field vector; **b** variation of the Strehl intensity ratio and the encircled energy for the E_x and E_y components as the mask order n_0 is increased

pupil plane. However since $A^x_{N,n}$ and $A^y_{N,n}$ are both real the weighting functions $a^x(u, \phi)$ and $a^y(u, \phi)$ are real.[4] Projecting back to the pupil plane is not a complex operation and hence the field in the pupil plane is also real. It is thus apparent that the field in the pupil plane is linearly polarised and only binary phase modulation is necessary. Apodisation is necessary as can be seen by considering

$$g(u, \phi) = \left(|\widetilde{E}_x|^2 + |\widetilde{E}_y|^2\right)^{1/2} = \frac{\sqrt{2 - u^2(1 - \sin 2\phi)}}{2(1 - u^2)} \sum_{n=0}^{\infty} A^x_{0,n} \Phi_{0,n}(c, u), \quad (6.28)$$

where a renormalisation of the incident field $\widetilde{\mathbf{E}}$ is required to ensure the mask is passive.

Practically, the series in Eqs. (6.27) must be truncated at say $n = n_0$, meaning the pupil and focal plane field distributions will differ from the ideal case in a way that is dependent on the truncation point. Figure 6.7a represents the required pupil plane distribution for $n_0 = 1$ whilst Fig. 6.8 shows the corresponding optical distribution in the focal plane. The shown distributions were calculated assuming NA = 0.966 and a value of $c = 4$, corresponding to a field of view in the focal plane approximately the size of the Airy disc.

From Fig. 6.8 it can be seen that there has been a resolution gain in the E_x and E_y distributions as compared to a clear aperture with uniform illumination, however there is little gain in the intensity focal spot. This again arises from a redistribution of energy to the unconstrained field component E_z which is then dominant in the final intensity profile. Furthermore, due to the presence of the apodising mask this

---

[4] The factor of i in Eq. (6.26) represents a global phase and can safely be ignored.

6.3 Inversion of the Debye-Wolf Diffraction Integral

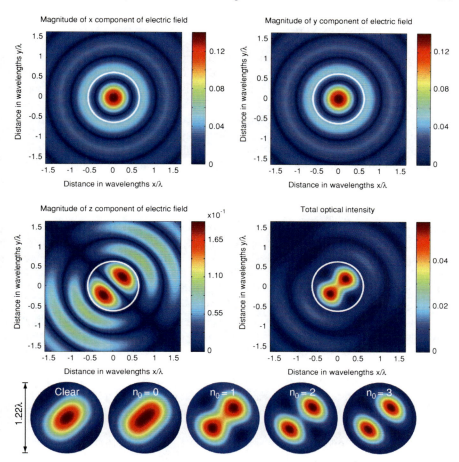

Fig. 6.8 Optical distribution in the focal plane for an apodised and polarisation structured beam with truncation point $n_0 = 1$ (*top*) for NA $= 0.966$ and $c = 4$. *White circles* again denote the extent of the Airy disc. Variation of the central intensity focal spot over the Airy disc as mask order n_0 is increased (*bottom*). Note the intensity scales differ with each plot, but have been equalised for easy comparison

arrangement also has a low optical efficiency, as can be seen in the plot of the Strehl intensity ratio (defined in [4]), shown in Fig. 6.7b, as a function of the truncation order n_0. At high n_0 this quantity loses its meaning however since the central peak essentially vanishes with respect to the large sidelobes, as would be expected from the discussion in Sect. 6.3.1.3. The performance of this particular superresolution setup consequently worsens as n_0 is increased.

6.3.2.2 Single Molecule Detection

As a second example consider trying to determine the orientation of a single fluorescent molecule. This often entails the use of a high NA optical system which provides the better resolution needed to select individual fluorophores. Many existing methods are limited to determination of the transverse angle [20, 53] and as such it would be desirable to couple light into the transverse orientation efficiently so as to improve the signal to noise ratio.[5] A fluorophore, modelled as a fixed electric dipole of moment **p**, illuminated by a field **E** re-radiates light as if it had an effective dipole moment proportional to $|\mathbf{p} \cdot \mathbf{E}|$. Efficient coupling thus entails minimising the longitudinal component of the focused field. Inverting a field specification of $E_z = 0$ gives a zero strength vector, meaning that such a specification cannot be achieved via apodisation or phase masks. However, a beam with a non-uniform polarisation distribution can be used. There exist numerous methods to produce these so-called vector beams including: modification of laser cavities, via introduction of polarisation sensitive components such that only modes with the desired polarisation structure can lase [34]; interferometric methods, which superpose orthogonal polarisation states with appropriate phase and intensity profiles [55]; and subwavelength gratings, which act as uniaxial crystals whose structure determines the birefringence [3].

Although these methods are typically used to generate radially and azimuthally polarised vector beams, it is possible to make more arbitrary vector beams by similar methods [59] or alternatively by using computer generated holograms and SLMs [35]. Of these, [35] is perhaps the most versatile being capable of dynamic modulation. Due to the pixelated nature of the SLMs however undesired diffraction effects can be introduced and complex algorithms are required. Toussaint et al. [59] avoided these issues albeit at the cost of reduced light throughput and complexity of the required optical setup.

In the ideal non-pixelated case the weighting function appropriate to a vector beam input is given by Eq. (6.13) with $g(u, \phi) = 1$ and $\Psi(u, \phi) = 0$. It has been observed that azimuthally polarised light, when focused, has a very weak longitudinal component [67]. Results from inversion agree with this observation as shown in Fig. 6.9, however a true azimuthal pattern is not seen due to an angular ambiguity in the inversion, meaning one half of the pattern is rotated by 180°. If the calculated polarisation structure is re-input into the forward focusing problem the maximum value of the longitudinal component is of order 10^{-17}; a number most likely attributable to numerical noise and inversion hence gives a suitable solution.

6.3.2.3 Extended Depth of Field

As a final example, extension of the depth of field in imaging systems is treated. In its most basic form extension of the depth of field can be considered as a problem of reducing the intolerance to defocus as judged by some pre-agreed figure of merit.

---

[5] A technique, developed by the author and colleagues, capable of measuring the longitudinal orientation is presented in Chap. 8.

6.3 Inversion of the Debye-Wolf Diffraction Integral

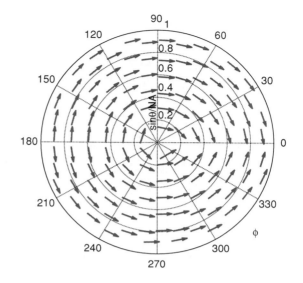

Fig. 6.9 Polarisation structure of illuminating beam required to give zero longitudinal field component in the focus of a high aperture lens as found by inversion (NA = 0.966, $c = 20$). *Arrows* indicate plane of polarisation of field at each point

Extended depth of field (EDF) in imaging systems has been considered by a number of researchers and engineers since it can become an important issue when imaging three dimensional objects and for design tolerances in optical systems. By far the most commonplace technique of extending the depth of field in an imaging system is by means of pupil plane engineering [10, 38, 63]. Other techniques also exist, including axial scanning and hybrid systems employing post-detection signal processing [43, 49], however a discussion of such methods will not be given here. Instead, a numerical example is given in which the incident beam is assumed to be uniformly x polarised. Consequently only the E_x field component contributes to the axial behaviour which is thus specified as

$$E_x(0, 0, z) = E_0 \operatorname{rect}\left(\frac{z}{w}\right), \quad (6.29)$$

where E_0 is a constant and w denotes the half width of the rect function. On axis Eq. (6.18) reduces to

$$\sum_{m=-\infty}^{\infty} J_m(kz) A^x_{m,0,n} = \frac{i}{kf}\left(\frac{u_\alpha}{k\rho_{\max}}\right) \frac{(-1)^n}{\lambda_{0,n}^{-1/2}} B^x_{0,n}, \quad (6.30)$$

since $\Phi_{|N|,n}(0) = 0$ for $N \neq 0$, i.e. only $N = 0$ orders contribute on axis.

Using Eqs. (6.17), (6.29) and (6.30) it is possible to numerically optimise the coefficients to find a good solution to the problem. One method of doing this is that of simulated annealing [28] in which random steps are taken with a probability that depends on a control parameter T which is slowly reduced. In simulated annealing, a loss function is defined which is analogous to the energy in an annealing process.

For the current example this was taken as the Hilbert angle ψ_H as defined by

$$\cos\psi_H = \frac{\left\langle |E_x(0,0,z)|^2, |E_x^{\text{opt}}(0,0,z)|^2 \right\rangle}{\||E_x(0,0,z)|^2\|^{1/2} \, \||E_x^{\text{opt}}(0,0,z)|^2\|^{1/2}}, \tag{6.31}$$

where

$$\left\langle |E_x(0,0,z)|^2, |E_x^{\text{opt}}(0,0,z)|^2 \right\rangle = \int_{-\infty}^{\infty} |E_x(0,0,z)|^2 |E_x^{\text{opt}}(0,0,z)|^2 dz, \tag{6.32}$$

and

$$\||E_x(0,0,z)|^2\| = \int_{-\infty}^{\infty} |E_x(0,0,z)|^4 dz. \tag{6.33}$$

The Hilbert angle is a measure of the similarity between the shape of the desired and optimised distributions $E_x(0,0,z)$ and $E_x^{\text{opt}}(0,0,z)$ respectively [49] ranging from 0 if they are identical, to $\pi/2$ if they are orthogonal.[6] Suitable truncation points for termination of the series in Eq. (6.30) can be determined as discussed in Sect. 6.2.2 so as to ensure convergence of the field expansion and was found to be $m_0 = 42$ for $w = 3\lambda$. Rejecting eigenvalues smaller than 10^{-5}, so as to limit noise amplification gave $n_0 = 9$. The resulting axial intensity profile as found from the 850 optimised coefficients is shown in Fig. 6.10a as compared to the desired rect function. The minimum Hilbert angle found was approximately $\frac{7\pi}{200}$.

The corresponding apodisation mask required to produce the optimised axial behaviour is shown in Fig. 6.10b and is very similar in form to a sinc mask as would be expected from McCutchen's theorem [32, 38]. Significant energy is however contained in the sidelobe structure (see Fig. 6.10c) and a fuller treatment may hence also incorporate a constraint of the sidelobe height by including additional terms in the loss function.

6.4 Polarisation Microscopy

6.4.1 Fisher Information in Microscopy

Optical microscopes are constructed so as to collect some portion of the field (or intensity) distribution originating from a sample object, and to form a magnified image of said field (or intensity) distribution on a scale of more practical use to an observer, be it human or machine. This rather vague description, does however obscure the variety of possible operational modalities of an optical microscope, which

---

[6] Whilst Eq. (6.31) strictly only compares the shape of the desired and optimised intensity profiles with no regard to the phase distribution, this is of little consequence to the results presented since only the intensity profile is deemed of any importance here.

6.4 Polarisation Microscopy

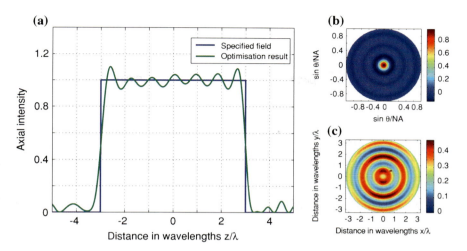

Fig. 6.10 a Comparison between the desired axial field profile and that found from a simulated annealing optimisation algorithm; **b** apodisation mask required to produce the optimised distribution (NA = 0.966); **c** resulting intensity distribution in the focal plane showing sidelobe pattern

arise from different configurations of the illumination and collection optics. Specifically reference to conventional and confocal microscopes are commonly found in the literature, e.g. [65]. In conventional microscopy a large area of a sample is illuminated and then imaged. Each point on the object is imaged in parallel and hence if a CCD were placed in the image plane of the objective lens a whole image could be obtained in a single instant. Alternatively a point detector could be used to build up the image pixel by pixel by scanning the position of the detector. Type I scanning microscopes, which operate on this principle, have been shown to possess the same imaging properties as a conventional microscope [64]. Confocal microscopes [47] on the other hand illuminate only a restricted portion of the object and then image only that region onto a point detector in the image plane of the system. Images are constructed pixelwise via synchronous scanning of the position of the illuminating point source and the point detector (although often it is more practical to scan the object).

Both type I scanning and confocal microscopes will record an intensity dependent on the domain from which light originates on the object Ω_{ob} and the extent of the detector Ω_{im}, as expressed by the integral

$$D(\Omega_{im}, \Omega_{ob}) = \iint_{\Omega_{im}} D_{im}(\rho, \Omega_{ob}) d\rho, \qquad (6.34)$$

where $D_{im}(\rho, \Omega_{ob})$ is the intensity at a point ρ in the detector/image plane due to light originating from Ω_{ob}.

Assuming the detector to be corrupted by classical shot noise the results of Sect. 5.2.3.1 can be used whereby the FIM matrix associated with estimation of the parameter vector $\mathbf{w}$ from a set of N_D intensity measurements $\mathbf{D} = (D_1, D_2, \ldots, D_{N_D})$ taken at a single point, is

$$\mathbb{J}_{\mathbf{w}}(\Omega_{\text{im}}, \Omega_{\text{ob}}) = \frac{\partial \mathbf{D}^\dagger}{\partial \mathbf{w}} \mathbb{J}_{\mathbf{D}} \frac{\partial \mathbf{D}}{\partial \mathbf{w}}, \tag{6.35}$$

where

$$\mathbb{J}_{\mathbf{D}} = \frac{1}{h\nu_0} \text{diag}\left[\frac{1}{D_i}\right] \tag{6.36}$$

and a potential background count has been neglected.

Scanning over the object, or image (or both), however implies that multiple measurements are taken, a fact which has not yet been accounted for. Since Fisher information is additive for independent measurements [16] the total FIM, $\mathbb{J}_T$, can be found by summing that obtained from each measurement, i.e.

$$\mathbb{J}_T = \sum_k \mathbb{J}_{\mathbf{w}}(\Omega_{\text{im}}^k, \Omega_{\text{ob}}^k), \tag{6.37}$$

where Ω_{im}^k and Ω_{ob}^k denote the respective domains for each scan point.[7]

Given the possibility of multiple measurement configurations it is important to consider the relative performance of each of them. As such a brief digression is made to prove the greater informational capabilities of imaging measurements as compared to a single, spatially averaged measurement taken over the same domain in the detection plane. For clarity, the results to follow are expressed in terms of a single intensity measurement and a single, real, scalar parameter w, however extension to the vector case will be discussed afterwards.

Consider first a wide area detector, which gives an output signal proportional to the integrated intensity over the detector area Ω_{im}. The associated Fisher information (using Eqs. (6.34) and (6.35) and neglecting a factor of $h\nu_0$) is given by

$$\begin{aligned}
J_{WA} &= \frac{1}{\iint_{\Omega_{\text{im}}} D_{\text{im}}(\rho, \Omega_{\text{ob}}) d\rho} \frac{\partial}{\partial w}\left[\iint_{\Omega_{\text{im}}} D_{\text{im}}(\rho, \Omega_{\text{ob}}) d\rho\right] \\
&\quad \times \frac{\partial}{\partial w}\left[\iint_{\Omega_{\text{im}}} D_{\text{im}}(\rho, \Omega_{\text{ob}}) d\rho\right], \\
&= \frac{1}{\iint_{\Omega_{\text{im}}} D_{\text{im}}(\rho, \Omega_{\text{ob}}) d\rho} \left|\iint_{\Omega_{\text{im}}} \frac{\partial D_{\text{im}}(\rho, \Omega_{\text{ob}})}{\partial w} d\rho\right|^2, \tag{6.38}
\end{aligned}$$

---

[7] Measurements from different positions can be stacked into a vector format and Eq. (6.35) used, however the assumption that the noise at each measurement position is independent allows simplification to Eq. (6.37), such that the dimensions of $\mathbf{D}$ are greatly reduced.

6.4 Polarisation Microscopy

if the region of integration does not depend on the parameter w. The equivalent result for an imaging arrangement, assuming continuous scanning such that the sum of Eq. (6.37) can be replaced by an integral, is given by

$$J_{IM} = \iint_{\Omega_{im}} \frac{1}{D_{im}(\rho, \Omega_{ob})} \left| \frac{\partial D_{im}(\rho, \Omega_{ob})}{\partial w} \right|^2 d\rho. \qquad (6.39)$$

Noting that optical intensity is a positive quantity, the Cauchy-Schwarz inequality, can be applied to Eq. (6.38) to give[8]

$$J_{WA} \leq \iint_{\Omega_{im}} \frac{1}{D_{im}(\rho, \Omega_{ob})} \left| \frac{\partial D_{im}(\rho, \Omega_{ob})}{\partial w} \right|^2 d\rho,$$

$$\leq J_{IM}, \qquad (6.40)$$

with equality when the intensity distribution is uniform over the detection plane. The inequality of Eq. (6.40) confirms the expectation that imaging an object field provides more information than a single, albeit spatially extended, measurement irrespective of the image formation process, due to the inherent averaging performed in the latter.

This result can also be extended to include multiple intensity measurements and vector parameters $\mathbf{w}$. Such an extension hence yields a matrix inequality $\mathbb{J}_{WA} \leq \mathbb{J}_{IM}$, where again the inequality implies that the difference matrix $\mathbb{J}_{IM} - \mathbb{J}_{WA}$ is positive semidefinite. Proof of this result centers on the positive definite nature of FIMs, from which it follows that $\mathbb{J}_{WA} \leq \mathbb{J}_{IM}$ holds if $\text{tr}(\mathbb{J}_{WA}) \leq \text{tr}(\mathbb{J}_{IM})$, a result which follows by applying the derivation above to each diagonal term of the FIM $\mathbb{J}_{WA}$ individually.

6.4.2 Examples

Electric and magnetic dipoles play a pivotal role in vectorial imaging. For example the vectorial Green's tensors are related to dipolar sources [61]. Furthermore, electric

---

[8] The Cauchy-Schwarz inequality reads

$$\left| \iint f(\rho)g(\rho) d\rho \right|^2 \leq \iint |f(\rho)|^2 d\rho \iint |g(\rho)|^2 d\rho,$$

which, with the substitutions

$$f(\rho) = \frac{1}{\sqrt{D_{im}(\rho, \Omega_{ob})}} \frac{\partial D_{im}(\rho, \Omega_{ob})}{\partial w} \quad \text{and} \quad g(\rho) = \sqrt{D_{im}(\rho, \Omega_{ob})},$$

yields

$$\left| \iint_{\Omega_{im}} \frac{\partial D_{im}(\rho, \Omega_{ob})}{\partial w} d\rho \right|^2 \leq \iint_{\Omega_{im}} \frac{1}{D_{im}(\rho, \Omega_{ob})} \left| \frac{\partial D_{im}(\rho, \Omega_{ob})}{\partial w} \right|^2 d\rho \iint_{\Omega_{im}} D_{im}(\rho, \Omega_{ob}) d\rho$$

thus leading to Eq. (6.40).

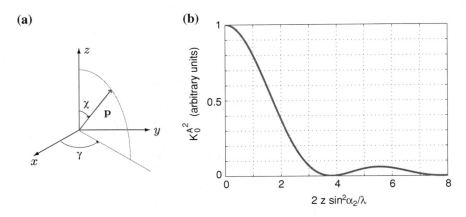

Fig. 6.11 a An electric dipole of moment **p** has a transverse and longitudinal orientation defined by the angles γ and χ respectively; **b** Variation of $K_0^{A^2}$ with defocus distance z_2, reflecting the reduction of Fisher information J_γ with defocus, when using a crossed polariser polarimeter

dipole emitters represent a good model for single molecules which are currently receiving much attention in the literature, as will be further discussed in Chap. 8. Due to their importance, the preceding theory will now be applied to the problem of imaging a single electric dipole. Although the theory for imaging magnetic dipoles is very similar it is omitted here for brevity. Estimation of the position of a dipole can be performed without resorting to polarisation based measurements and has, for example, been considered in [37, 44]. Instead, the determination of the transverse orientation of the dipole, as described by an angle γ (see Fig. 6.11a) is considered in this section. Limitation to measurement of the transverse orientation is made (i.e. χ is assumed to be $\pi/2$) since optical microscopes are unable to measure the longitudinal component easily. Further attention will however be given to the problem of determination of the longitudinal component of the dipole moment in Chap. 8.

Dipolar crosstalk will also be investigated by considering the reduction in Fisher information when a second dipole is introduced to the object plane. This example can prove insightful when considering the multiplexed optical data storage technique discussed in Chap. 7 by providing a first order approximation to crosstalk that can occur between neighbouring data pits on an optical disc. In all examples Poisson noise will be assumed as this is often present in single molecules experiments.

6.4.2.1 Transverse Dipole Orientation

Before calculating the Fisher information pertinent to inference of the orientation of a single dipole, it is first necessary to calculate the associated intensity distribution in the detector plane. To do so the simple transmission geometry of Fig. 6.12 is assumed, where the collector and detector lens have numerical apertures $NA_1 = \sin\alpha_1$ and

6.4 Polarisation Microscopy

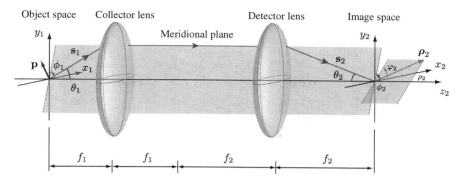

Fig. 6.12 Simple $4f$ telecentric imaging setup used to image a single dipole. Positions in the object and image plane are defined by the position vectors ρ_1 and ρ_2 respectively, whilst positions on the reference spheres associated with the collector and detector lens (assumed aplanatic and with numerical apertures $NA_1 = \sin \alpha_1$ and $NA_2 = \sin \alpha_2$) are defined by the coordinates (θ_1, ϕ_1) and (θ_2, ϕ_2). Ray directions in the respective spaces are described by the normalised wavevectors $\mathbf{s}_1$ and $\mathbf{s}_2$

$NA_2 = \sin \alpha_2$ respectively. The dipole, with moment $\mathbf{p} = (p_x, p_y, p_z)$ is assumed to lie on the optical axis in the front focal plane of the first lens, whilst the detector is placed in the focal plane of the second lens.

Owing to the telecentric nature of a marked proportion of microscope objective lenses the vectorial ray tracing formulation presented in Chap. 4 can again be applied. The field distribution arising from the radiating dipole source on a sphere located in the pupil of the first lens at infinity, can be found by considering the far field distribution of an electric dipole,[9] as given by

$$\mathbf{e}_1(\theta_1, \phi_1) = \mathbf{s}_1 \times (\mathbf{s}_1 \times \mathbf{p}) \frac{\exp(ik\rho_1)}{\rho_1}, \qquad (6.41)$$

where $\mathbf{s}_1 = (\sin \theta_1 \cos \phi_1, \sin \theta_1 \sin \phi_1, \cos \theta_1)^T$. The exponential term is however constant over a sphere and can hence be dropped. Applying the inverse of Eq. (47), whereby $\widetilde{\mathbf{E}}_1(\theta_1, \phi_1) = \mathbb{Q}^{-1}(\theta_1, \phi_1) \cdot \mathbf{e}(\theta_1, \phi_1)$, gives the collimated field after the collector lens as

$$\widetilde{\mathbf{E}}_1(\theta_1, \phi_1) = \frac{1}{2\sqrt{\cos\theta_1}} \begin{pmatrix} (q_1 + q_2 \cos 2\phi_1)p_x + q_2 \sin 2\phi_1 p_y - q_3 \cos \phi_1 p_z \\ q_2 \sin 2\phi_1 p_x + (q_1 - q_2 \cos 2\phi_1)p_y - q_3 \sin \phi_1 p_z \\ 0 \end{pmatrix}, \qquad (6.42)$$

where $q_i(\theta)$ are given by Eqs. (4.49).

Refocusing of the collimated field gives rise to a field, $\mathbf{E}_2(\rho_2 = (\rho_2, \varphi_2, z_2))$, in the image plane as can be found by evaluation of the Debye-Wolf diffraction

---

[9] For magnetic dipoles the far field distribution takes the form $(\mathbf{s}_1 \times \mathbf{p}) \exp(ik\rho_1)/\rho_1$.

integral. The calculations are very similar to those presented in Sect. 4.5.1 and are therefore omitted here for brevity. Full exposition can, however, be found in [33, 56] where it is shown that

$$\mathbf{E}_2(\rho_2) = \begin{pmatrix} p_x(K_0^A + K_2^A \cos 2\varphi_2) + p_y K_2^A \sin 2\varphi_2 + 2ip_z K_1^A \cos \varphi_2 \\ p_x K_2^A \sin 2\varphi_2 + p_y(K_0^A - K_2^A \cos 2\varphi_2) + 2ip_z K_1^A \sin \varphi_2 \\ -2i(p_x \cos \varphi_2 + p_y \sin \varphi_2)K_1^B - 2p_z K_0^B \end{pmatrix},$$
(6.43)

where

$$K_0^A = \int_0^{\alpha_2} \sqrt{\frac{\cos \theta_2}{\cos \theta_1}} \sin \theta_2 (1 + \cos \theta_1 \cos \theta_2) J_0(k\rho_2 \sin \theta_2) \exp[ikz_2 \cos \theta_2] d\theta_2$$
(6.44a)

$$K_0^B = \int_0^{\alpha_2} \sqrt{\frac{\cos \theta_2}{\cos \theta_1}} \sin^2 \theta_2 \sin \theta_1 J_0(k\rho_2 \sin \theta_2) \exp[ikz_2 \cos \theta_2] d\theta_2$$
(6.44b)

$$K_1^A = \int_0^{\alpha_2} \sqrt{\frac{\cos \theta_2}{\cos \theta_1}} \sin \theta_2 \sin \theta_1 \cos \theta_2 J_1(k\rho_2 \sin \theta_2) \exp[ikz_2 \cos \theta_2] d\theta_2$$
(6.44c)

$$K_1^B = \int_0^{\alpha_2} \sqrt{\frac{\cos \theta_2}{\cos \theta_1}} \sin^2 \theta_2 \cos \theta_1 J_1(k\rho_2 \sin \theta_2) \exp[ikz_2 \cos \theta_2] d\theta_2$$
(6.44d)

$$K_2^A = \int_0^{\alpha_2} \sqrt{\frac{\cos \theta_2}{\cos \theta_1}} \sin \theta_2 (1 - \cos \theta_1 \cos \theta_2) J_2(k\rho_2 \sin \theta_2) \exp[ikz_2 \cos \theta_2] d\theta_2.$$
(6.44e)

These K integrals can be evaluated numerically when complemented with the aplanatic condition $\sin \theta_1 = \beta \sin \theta_2$, where $\beta = f_2/f_1$ is the magnification of the imaging system as determined by the ratio of the focal lengths f_1 and f_2 of the two lenses.

Restricting now to on-axis detection with a point polarimetric detector, Eq. (6.43) reduces to $\mathbf{E}_2(\rho_2 = 0) = (K_0^A p_x, K_0^A p_y, 0)^T$. Equivalently the on-axis, in-focus Stokes vector is given by $\mathbf{S}(\rho_2 = 0) = (K_0^{A^2} p_0^2, K_0^{A^2} p_0^2 \cos 2\gamma, K_0^{A^2} p_0^2 \sin 2\gamma, 0)^T$, where the substitutions $p_x = p_0 \cos \gamma$, $p_y = p_0 \sin \gamma$ and $p_z = 0$ have also been made. Upon measurement using a polarimeter a set of N_D measured intensities, $\mathbf{D} = \mathbb{V}\mathbb{T}\mathbf{S}$, results. Unfortunately, inference of the dipole orientation requires the additional estimation of the intensity $S_0 = K_0^{A^2} p_0^2$. Accordingly the FIM for estimation of the parameter vector $\mathbf{w} = (S_0, \gamma)^T$ is given by

$$\mathbf{J_w} = \frac{\partial \mathbf{S}}{\partial \mathbf{w}}^T \mathbb{T}^T \mathbb{V}^T \mathbf{J}_D \mathbb{V}\mathbb{T} \frac{\partial \mathbf{S}}{\partial \mathbf{w}},$$
(6.45)

6.4 Polarisation Microscopy

where

$$\frac{\partial \mathbf{S}}{\partial \mathbf{w}} = \begin{pmatrix} 1 & 0 \\ \cos 2\gamma & -2S_0 \sin 2\gamma \\ \sin 2\gamma & 2S_0 \cos 2\gamma \\ 0 & 0 \end{pmatrix}. \tag{6.46}$$

As described in Sect. 3.3.1, the intensity can be treated as a nuisance parameter and the reduced Fisher information pertaining to estimation of the dipole orientation can be found. Bearing in mind the form of $\mathbf{E}_2(\rho_2 = \mathbf{0})$, perhaps the most intuitive approach to measurement of the dipole orientation is to use a DOAP comprising a pair of polarisers with corresponding instrument matrix[10] given by

$$\mathbb{T} = \frac{1}{2} \begin{pmatrix} 1 & \cos 2\vartheta_1 & \sin 2\vartheta_1 & 0 \\ 1 & \cos 2\vartheta_2 & \sin 2\vartheta_2 & 0 \end{pmatrix}, \tag{6.47}$$

where ϑ_i define the azimuthal angles of the measurement states in Poincaré space. This approach is also justifiable since, by the model and assumptions described above, it is known a priori that the light incident onto the detector is both fully and linearly polarised, such that the results of Sect. 5.3.1.3 are applicable. Following the discussion given in Sect. 5.3, it will also be assumed that $\mathbb{V} = \mathbb{I}/N_D$, i.e. that light is distributed equally among each measurement arm of the DOAP. From Eqs. (6.36) and (6.45–6.47) the Fisher information J_γ can then be explicitly found, neglecting for the moment a possible background intensity, viz.

$$J_\gamma = \frac{2S_0}{h\nu_0} \frac{\sin^2(\vartheta_1 - \vartheta_2)}{\cos^2(\gamma - \vartheta_1) + \cos^2(\gamma - \vartheta_2)}. \tag{6.48}$$

Maximum Fisher information is thus achieved when $\vartheta_1 = \vartheta_2 + \pi/2$, i.e. when the polarisers are crossed, whereupon Eq. (6.48) simplifies to $J_\gamma = J_\gamma^{\max} = 2S_0/h\nu_0$ for all dipole orientations.

For comparison with the following section it is useful to consider the Fisher information J_γ when S_0 is assumed known a priori. In particular for two arbitrarily oriented polarisers the resulting expression is

$$J_\gamma = J_\gamma^{\max} \left[\sin^2(\gamma - \vartheta_1) + \sin^2(\gamma - \vartheta_2) \right], \tag{6.49}$$

which again reduces to $J_\gamma = J_\gamma^{\max}$ for all dipole orientations, when the polarisers are crossed.[11] This behaviour has been shown in Fig. 6.13a in conjunction with the results obtained when the DOAPs introduced in Chap. 5 are used to measure

---

[10] Two measurements are required such that the resulting set of linear equations are well conditioned (neglecting an ambiguity as to which quadrant the dipole lies in), when noise is absent.

[11] Whilst the notation $J_\gamma^{\max}$ will be retained, it should be observed that when S_0 is known a priori, the maximum Fisher information is $2J_\gamma^{\max}$.

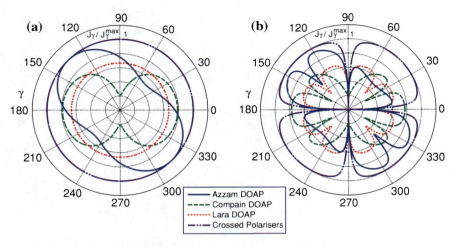

Fig. 6.13 Polar plots, showing the calculated Fisher information $J_\gamma(\gamma)/J_\gamma^{\max}$, for estimation of the transverse orientation γ of a single electric dipole, for zero background intensity (**a**) and for $S_0/D_b = 10^3$ (**b**). Informational dips are introduced in the presence of a background count. The longitudinal angle χ is assumed to be $\pi/2$ and the dipole assumed to be imaged using a simple $4f$ system with $100\times$ magnification, in which the collector lens has a numerical aperture of 0.95 and measured by an on-axis DOAP configured as in Chap. 5

the state of polarisation in focus. The improved, and γ independent performance of the crossed polariser arrangement is evident. This behaviour arises since the DOAPs of Chap. 5 are designed for measurement of the ellipticity, in addition to, the azimuthal angle of a state of polarisation. Furthermore the DOAPs of Chap. 5 possess more detection arms (see Sect. 5.2.4), however can fully determine which quadrant γ lies in.

Defocus in the imaging system, or equivalently an axial shift in the position of the dipole, can also be easily characterised since the axial coordinate appears only in the K integrals.[12] The variation of $K_0^{A^2}$, and hence J_γ as a function of defocus distance z_2 is shown in Fig. 6.11b. Oscillatory behaviour is exhibited, however best performance is seen within the depth of focus of the imaging system as would be expected.

Presence of a background intensity during the measurement process however produces some interesting behaviour and breaks the orientation independence of the Fisher information as shall now be considered. Maintaining the assumption that S_0 is known a priori for simplicity, when a background intensity D_b is incorporated (assumed the same on each detector), Eq. (6.49) becomes

---

[12] The K integrals of Eqs. (6.44) accommodate a defocus of the detector in the image plane. If however the dipole suffers an axial shift, z_{dp} the exponential term $\exp[ikz_2\cos\theta_2]$ must be modified to $\exp[ik(z_2\cos\theta_2 - z_{dp}\cos\theta_1)]$ (see [33]).

6.4 Polarisation Microscopy

$$J_\gamma = J_\gamma^{\max} \left[\frac{S_0 \sin^2(\gamma - \vartheta_1) \cos^2(\gamma - \vartheta_1)}{S_0 \cos^2(\gamma - \vartheta_1) + D_b} + \frac{S_0 \sin^2(\gamma - \vartheta_2) \cos^2(\gamma - \vartheta_2)}{S_0 \cos^2(\gamma - \vartheta_2) + D_b} \right]. \tag{6.50}$$

Assuming the polarisers are crossed at angles of $0°$ and $90°$ respectively, gives

$$J_\gamma = J_\gamma^{\max} S_0 \sin^2 \gamma \cos^2 \gamma \left[\frac{1}{S_0 \cos^2 \gamma + D_b} + \frac{1}{S_0 \sin^2 \gamma + D_b} \right]. \tag{6.51}$$

Whilst in the absence of a background count, a constant Fisher information was seen, if the limits $\gamma \to 0°$ and $\gamma \to 90°$ are taken, Eq. (6.51) yields zero Fisher information, as shown in Fig. 6.13b (assuming the ratio $S_0/D_b = 10^3$). Dipole orientations which differ from these critical orientations however still approximately exhibit the γ invariant behaviour of J_γ. Furthermore, it is found that as the ratio S_0/D_b decreases so the width of the information dip around the critical angles increases to such an extent that for large D_b, $J_\gamma \propto \sin 2\gamma$. For reference, it is noted that this is of the same functional form as when Gaussian noise is assumed.

Explanation of this behaviour can be found by first considering inference of the orientation of a dipole using a single polariser only. The Fisher information in this case is given by a single term of Eq. (6.51). Inference from a single polariser centers around Malus' law[13] with a zero intensity corresponding to a dipole orientation perpendicular to the polariser, whilst parallel orientations give a maximum intensity. The scaling of the variance of Poisson noise with the mean (see Table 2.1), however, automatically implies that a zero intensity suffers no noise, such that it can be identified exactly, whilst the converse holds for a maximum intensity. Furthermore, Malus' law means that small changes in the orientation of a dipole aligned close to these critical orientations, give rise to very small changes in the measured intensity (mathematically $\partial D/\partial \gamma = 0$, when the dipole and polariser are co- or perpendicularly oriented). Due to low noise levels for a perpendicularly oriented dipole, small changes in its orientation can be distinguished. If, however, a dipole is originally parallel to the polariser, the small intensity changes associated with small rotations are lost in the noise.

Fisher information, being a local measure of information, quantifies the sensitivity of the recorded intensities on the dipole orientation and hence a single polariser oriented at $0°$ yields zero Fisher information with regards to a dipole oriented at $\gamma = 0°$, however provides maximum information for $\gamma = 90°$. Thus if two crossed polarisers are used, a reduction in Fisher information from a measurement in one is counterbalanced by a gain in the other, producing an orientation independent J_γ.

The argument above however only holds when there is no additional noise present in the system. Additional noise, arising from say a background count, has the adverse effect of masking small changes in intensity arising from small angular deflections

---

[13] Malus' law states that the intensity transmitted through a polariser with transmission axis at ϑ when illuminated with light, linearly polarised at an angle γ, is given by $D \propto \cos^2(\vartheta - \gamma)$.

of a perpendicularly oriented dipole, in a similar manner to how small changes of a co-oriented dipole are lost in the inherent Poisson noise.

Larger dipole deflections give rise to larger intensity changes in the detector and thus when measuring dipole orientations which differ significantly from the critical angles, the background count becomes less relevant and the background free behaviour is restored. The extent of angular deviations considered to be significantly far from the critical orientations is naturally set by the strength of the background noise, hence explaining the widening of the information dip as S_0/D_b decreases.

Informational dips such as that discussed for a pair of crossed polarisers can also be seen for alternative DOAP architectures as shown in Fig. 6.13b. Importantly, it should be noted that no dipole orientations give rise to complete loss of Fisher information, since each of these DOAP configurations possess additional arms in which a non-zero information is obtained, i.e. there is a redundancy in the measurements, for example the DOAP of Lara and Paterson (red dashed line) [29] possesses additional polarisers at $\pm 45°$. Polarisers at $\pm 45°$ also suffer an information dip for dipoles orientated at $\pm 45°$, however these dips are then reduced by the presence of the detection paths with crossed polarisers oriented at $0°$ and $90°$, which yield non-zero information, in a reciprocal manner.

6.4.2.2 Dipolar Crosstalk

In the preceding section the image of an on-axis dipole was calculated and used to quantify the accuracy achievable in inferring the dipole's orientation. Introduction of a second, extraneous, off-axis dipole in the object plane, however modifies the image plane distribution and thus can have consequences on any estimate of the first dipole's orientation, as shall now be analysed.

Consider first then the image of an off-axis dipole. Displacement of a dipole to an off-axis position $\rho_{dp} = (\rho_{dp}, \varphi_{dp}, 0)^T$ can be modelled by assuming shift invariance of the imaging system. Shift invariant imaging can be justified following the discussion in [33], where the displacement is shown to introduce an additional phase term into the Debye-Wolf diffraction integral, allowing the image of an off-axis dipole to be calculated using Eq. (6.43) whereby $\mathbf{E}_2^{\text{off-axis}}(\rho_2) = \mathbf{E}_2^{\text{on-axis}}(\rho_2 - \beta \rho_{dp})$. The image field of the jth dipole in this example ($j = 1$ or 2), with moment $\mathbf{p}_j = p_0(\cos\gamma_j, \sin\gamma_j, 0)^T$, can then be expressed in the form

$$\mathbf{E}^{dp_j} = p_0 \begin{pmatrix} a_{1j+} \cos\gamma_j + a_{2j} \sin\gamma_j \\ a_{2j} \cos\gamma_j + a_{1j-} \sin\gamma_j \\ a_{3j} i \cos\gamma_j + a_{4j} i \sin\gamma_j \end{pmatrix}, \qquad (6.52)$$

where

$$a_{1j\pm} = K_{0j}^A \pm K_{2j}^A \cos 2\Phi_j \qquad (6.53)$$

$$a_{2j} = K_{2j}^A \sin 2\Phi_j, \qquad (6.54)$$

6.4 Polarisation Microscopy

$$a_{3j} = -2K_{1j}^B \cos \Phi_j, \tag{6.55}$$

$$a_{4j} = -2K_{1j}^B \sin \Phi_j. \tag{6.56}$$

The index on the K integrals denotes the dependence on the radial coordinate $|\rho_2 - \beta\rho_{\text{dp}_j}|$ of the jth dipole, whilst $\Phi_j = \arctan[(y_2 - \beta y_{\text{dp}_j})/(x_2 - \beta x_{\text{dp}_j})]$.

The total intensity in the detector plane will however vary depending on the coherence properties of the light originating from the two dipoles. Two limiting cases will be considered here in which the light from the dipoles is either fully incoherent or fully coherent. The former situation may arise, for example, if imaging single fluorescent molecules, e.g. in fluorescence microscopy [24], where the inherently random nature of the excitation and re-emission process results in incoherent radiation, such that *intensities*, or equivalently Stokes parameters, add in the image plane. Alternatively, if imaging two dipoles induced by a coherent field, as may arise in the scattering of light from gold beads, or other small scatterers [56, 60], the dipoles radiate coherently, meaning *field vectors* sum in the image plane.

Detection is assumed to be performed by a scanning point DOAP comprising two polarisers oriented at $\vartheta_1 = 0$ and $\vartheta_2 = \pi/2$, i.e. horizontally and vertically crossed polarisers as described in the preceding section. Strictly, the presence of a non-zero longitudinal field component for off-axis dipoles, necessitates a full 3D treatment as discussed in Chap. 4, whereby generalised Stokes vectors become 9×1 vectors. The instrument matrix for the two crossed polariser scenario can then be shown to be given by

$$\mathbb{T} = \begin{pmatrix} \frac{2}{3} & \frac{1}{2} & 0 & 0 & 0 & 0 & 0 & -\frac{1}{2\sqrt{3}} \\ \frac{2}{3} & -\frac{1}{2} & 0 & 0 & 0 & 0 & 0 & -\frac{1}{2\sqrt{3}} \end{pmatrix}. \tag{6.57}$$

Accordingly, the detected intensity vector (as still found using $\mathbf{D} = \mathbb{V}\mathbb{T}\mathbf{S}$ and once more neglecting background readings) is given by

$$\mathbf{D}^{\text{inc}} = \frac{1}{2} \begin{pmatrix} |E_x^{\text{dp}_1}|^2 + |E_x^{\text{dp}_2}|^2 + |E_z^{\text{dp}_1}|^2 + |E_z^{\text{dp}_2}|^2 \\ |E_y^{\text{dp}_1}|^2 + |E_y^{\text{dp}_2}|^2 + |E_z^{\text{dp}_1}|^2 + |E_z^{\text{dp}_2}|^2 \end{pmatrix}, \tag{6.58}$$

for the incoherent dipole case, or alternatively by

$$\mathbf{D}^{\text{coh}} = \frac{1}{2} \begin{pmatrix} |E_x^{\text{dp}_1} + E_x^{\text{dp}_2}|^2 + |E_z^{\text{dp}_1} + E_z^{\text{dp}_2}|^2 \\ |E_y^{\text{dp}_1} + E_y^{\text{dp}_2}|^2 + |E_z^{\text{dp}_1} + E_z^{\text{dp}_2}|^2 \end{pmatrix}, \tag{6.59}$$

for coherently radiating dipoles.

To evaluate the extent of crosstalk between the dipoles when attempting to infer the orientation of one dipole in the presence of Poisson noise, the FIM is again calculated. The magnitude of the dipole moment p_0 is assumed to be known a priori for simplicity and hence so too is S_0, such that

$$J_{\gamma_1}^{\nu}(\gamma_1, \gamma_2) = \frac{1}{D_1^{\nu}}\left(\frac{\partial D_1^{\nu}}{\partial \gamma_1}\right)^2 + \frac{1}{D_2^{\nu}}\left(\frac{\partial D_2^{\nu}}{\partial \gamma_1}\right)^2 \qquad (6.60)$$

for $\nu =$ coh or inc, denoting the coherent and incoherent case respectively. Calculating $J_{\gamma_1}^{\nu}$ requires the derivatives

$$\frac{\partial D_1^{\text{inc}}}{\partial \gamma_j} = p_0 \Big[(a_{2j} \cos \gamma_j - a_{1j+} \sin \gamma_j)(a_{1j+} \cos \gamma_j + a_{2j} \sin \gamma_j)$$
$$+ (a_{4j} \cos \gamma_j - a_{3j} \sin \gamma_j)(a_{3j} \cos \gamma_j + a_{4j} \sin \gamma_j) \Big], \qquad (6.61)$$

$$\frac{\partial D_2^{\text{inc}}}{\partial \gamma_j} = p_0 \Big[(a_{2j} \cos \gamma_j + a_{1j-} \sin \gamma_j)(a_{1j-} \cos \gamma_j - a_{2j} \sin \gamma_j)$$
$$+ (a_{4j} \cos \gamma_j - a_{3j} \sin \gamma_j)(a_{3j} \cos \gamma_j + a_{4j} \sin \gamma_j) \Big], \qquad (6.62)$$

and

$$\frac{\partial D_1^{\text{coh}}}{\partial \gamma_j} = p_0 \Big[(a_{2j} \cos \gamma_j - a_{1j+} \sin \gamma_j)(a_{11+} \cos \gamma_1 + a_{12+} \cos \gamma_2$$
$$+ a_{21} \sin \gamma_1 + a_{22} \sin \gamma_2) + (a_{4j} \cos \gamma_j - a_{3j} \sin \gamma_j)$$
$$\times (a_{31} \cos \gamma_1 + a_{32} \cos \gamma_2 + a_{41} \sin \gamma_1 + a_{42} \sin \gamma_2) \Big], \qquad (6.63)$$

$$\frac{\partial D_2^{\text{coh}}}{\partial \gamma_j} = p_0 \Big[(a_{1j-} \cos \gamma_j - a_{2j} \sin \gamma_j)(a_{21} \cos \gamma_1 + a_{22} \cos \gamma_2$$
$$+ a_{11-} \sin \gamma_1 + a_{12-} \sin \gamma_2) + (a_{4j} \cos \gamma_j - a_{3j} \sin \gamma_j)$$
$$\times (a_{31} \cos \gamma_1 + a_{32} \cos \gamma_2 + a_{41} \sin \gamma_1 + a_{42} \sin \gamma_2) \Big]. \qquad (6.64)$$

Examining the incoherent dipole case first, it is noted that the derivative terms, $\partial D_i^{\text{inc}}/\partial \gamma_1$, are independent of γ_2 and ρ_{dp_2}, whilst the additive terms, $|E_i^{\text{dp}_2}|^2$, ($i = x, y, z$) in Eq. (6.58) are independent of γ_1 and ρ_{dp_1}. As such, the second dipole acts as a background source. For all positions and orientations of the second dipole with a non-zero intensity at the detection point, information dips akin to those discussed in the single dipole example above, are introduced into the performance characteristics of any estimator for $\gamma_1 = 0°$ or $90°$.

With respect to coherent dipoles, consider briefly the limiting case in which the second dipole is moved to infinity, i.e. $\rho_2 \to \infty$. The functional dependence of the K integrals of Eq. (6.44) implies that $\{a_{12\pm}, a_{22}, a_{32}, a_{42}\} \to 0$ as $\rho_2 \to \infty$. Under these circumstances

6.4 Polarisation Microscopy

$$\frac{\partial D_1^{coh}}{\partial \gamma_1} \rightarrow p_0 \Big[(a_{21} \cos \gamma_j - a_{11+} \sin \gamma_1)(a_{11+} \cos \gamma_1 + a_{21} \sin \gamma_1)$$
$$+ (a_{41} \cos \gamma_1 - a_{31} \sin \gamma_1)(a_{31} \cos \gamma_1 + a_{41} \sin \gamma_1) \Big] = \frac{\partial D_1^{inc}}{\partial \gamma_1},$$

$$\frac{\partial D_2^{coh}}{\partial \gamma_1} \rightarrow p_0 \Big[(a_{11-} \cos \gamma_1 - a_{21} \sin \gamma_1)(a_{21} \cos \gamma_1 + a_{11-} \sin \gamma_1)$$
$$+ (a_{41} \cos \gamma_1 - a_{31} \sin \gamma_1)(a_{31} \cos \gamma_1 + a_{41} \sin \gamma_1) \Big] = \frac{\partial D_2^{inc}}{\partial \gamma_1},$$

whereupon it is seen that coherent and incoherent cases exhibit the same behaviour, for large dipole separations.

Limiting forms of $J_{\gamma_1}^\nu$ can also be derived when the second dipole is at a finite distance from the first, which shall now be assumed to again be located on-axis. Of significance is the form of Eq. (6.60) when the $\mathbf{p}_1$ and $\mathbf{p}_2$ are parallel or perpendicular, since these configurations represent the two extreme configurations. Furthermore, the cases when $\mathbf{p}_1$ lies parallel or at 45° to the transmission axis of one of the analysing polarisers will be considered. Expressions for $J_{\gamma_1}^\nu(\gamma_1, \gamma_2)$ are hence now given when $\gamma_1 = 0$ for $\gamma_2 = 0$ and $\pi/2$ and also for $\gamma_1 = \pi/4$ and $\gamma_2 = \pm\pi/4$. For the incoherent case

$$J_{\gamma_1}^{inc}(0, 0) = J_{\gamma_1}^{inc}\left(0, \tfrac{\pi}{2}\right) = 0, \tag{6.65a}$$

$$J_{\gamma_1}^{inc}\left(\tfrac{\pi}{4}, \pm\tfrac{\pi}{4}\right) = \frac{J_\gamma^{max}}{2}\Bigg[\frac{a_{11+}^2}{a_{11+}^2 + (a_{12+} \pm a_{22})^2 + (a_{32} \pm a_{42})^2}$$
$$+ \frac{a_{11-}^2}{a_{11-}^2 + (a_{22} \pm a_{12-})^2 + (a_{32} \pm a_{42})^2}\Bigg]. \tag{6.65b}$$

where the expected information loss at $\gamma_1 = 0°$ and 90° is clearly evident. Coherent superposition of the radiated dipole fields yields

$$J_{\gamma_1}^{coh}(0, 0) = J_\gamma^{max} \frac{a_{22}^2}{a_{22}^2 + a_{32}^2}, \tag{6.66a}$$

$$J_{\gamma_1}^{coh}\left(0, \tfrac{\pi}{2}\right) = J_\gamma^{max} \frac{a_{12-}^2}{a_{12-}^2 + a_{42}^2}, \tag{6.66b}$$

$$J_{\gamma_1}^{coh}\left(\tfrac{\pi}{4}, \pm\tfrac{\pi}{4}\right) = \frac{J_\gamma^{max}}{2}\Bigg[\frac{(a_{11+} + a_{12+} \pm a_{22})^2}{(a_{11+} + a_{12+} \pm a_{22})^2 + (a_{32} \pm a_{42})^2}$$
$$+ \frac{(a_{22} + a_{11-} \pm a_{12-})^2}{(a_{22} + a_{11-} \pm a_{12-})^2 + (a_{32} \pm a_{42})^2}\Bigg]. \tag{6.66c}$$

Inspection of the denominators in Eqs. (6.66) reveals that the longitudinal field component arising in the image plane from the extraneous dipole, acts as a background source, a result which also holds for more general configurations (if the dipole of interest is located on-axis). Informational losses therefore can once more result. Since the second lens is however practically of low numerical aperture, this background term is small and the dips correspondingly narrow.

Dropping the longitudinal background term thus allows the interference effects for coherent dipoles to be considered more closely. By evaluation of Eq. (6.60) it can be shown that

$$J_{\gamma_1}^{\mathrm{coh}}(\gamma_1, \gamma_2) = \frac{2p_0^2 K_{01}^{A~2}}{h\nu_0} = J_\gamma^{\max}, \qquad (6.67)$$

for all dipole configurations. The presence of a second dipole is thus seen not to affect $J_{\gamma_1}^{\mathrm{coh}}$. That said a bias, in general dependent on γ_1, γ_2 and ρ_{dp_2}, is introduced into an estimator, with a resulting increase in the variance, and mean squared error, as per Eq. (3.27). As the interfering dipole is gradually moved to larger distances the magnitude of this bias decreases, so as to restore the single dipole results of Sect. 6.4.2.1.

Removal of the estimator bias could be approached by a reformulation of the problem as one of the joint estimation of (γ_1, γ_2), in which the orientation of the second dipole is treated as a nuisance parameter. To illustrate the general performance that can be expected, a Bayesian viewpoint is adopted in which the orientation of the second dipole is assumed to obey a uniform PDF, i.e. $f_{\Gamma_2}(\gamma_2) = 1/2\pi$ and the Fisher information for inference of γ_1 calculated as a function of dipole separation and γ_1. The second dipole is assumed to lie on the positive x axis in object space and the same imaging parameters assumed in Sect. 6.4.2.1 are used.

When considering the coherent, zero background results ($D_b = 0$) shown in Fig. 6.14a, it is also useful to consider Fig. 6.14b which shows the variation of $J_{\gamma_1}^{\mathrm{coh}}(\gamma_1)$ with dipole separation, $\rho_{\mathrm{dp}_2} = x_{\mathrm{dp}_2}$, for $\gamma_1 = 0°$ and $90°$ (solid lines). Plots of $K_{02}^A(\rho_2) \pm K_{22}^A(\rho_2)$ (normalised such that $K_{02}^A(0) = 1$) are also shown (dashed lines). These latter plots correspond to the variation of $E_x^{\mathrm{dp}_2}(\rho_2)$ and $E_y^{\mathrm{dp}_2}(\rho_2)$ respectively. Peaks in $J_{\gamma_1}^{\mathrm{coh}}(0)$ are seen to correspond to zeros in $E_x^{\mathrm{dp}_2}(\rho_2)$, whilst maxima in $J_{\gamma_1}(\pi/2)$ correspond to zeros in $E_y^{\mathrm{dp}_2}(\rho_2)$. Such a situation can be understood by first noting that an x-oriented dipole yields a zero intensity in D_2 when the second dipole is absent. The presence of a second dipole hence introduces noise into D_2 when $E_y^{\mathrm{dp}_2}(\rho_2) \neq 0$, thus destroying the ability of an observer to identify the zero signal from the first dipole, in an analogous manner to the formation of informational dips. The scenario is similar for a y-oriented dipole albeit the role of the detectors is reversed. A peak in $J_{\gamma_1}^{\mathrm{coh}}(0)$ also corresponds to a minimum in $J_{\gamma_1}^{\mathrm{coh}}(\pi/2)$ (and vice-versa), by similar arguments.

When an independent background count D_b is introduced, zero Fisher information is seen for x- and y-oriented dipoles regardless of the position or angle of the second dipole as shown in Fig. 6.14c, where the ratio $S_0/D_b = 10^2$ was assumed.

6.4 Polarisation Microscopy

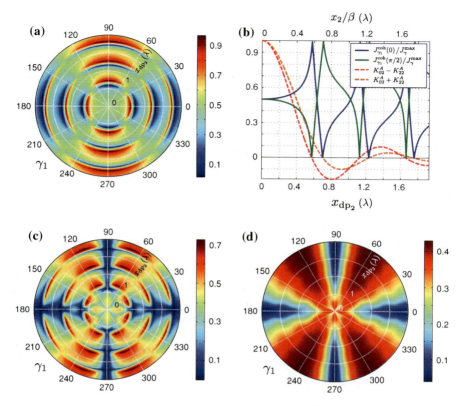

Fig. 6.14 a Calculated normalised Bayesian Fisher information $J_{\gamma_1}^{\text{coh}}$ for estimation of the orientation of an on-axis dipole in the presence of a second extraneous dipole located at $\rho_{\text{dp}_2} = (x_{\text{dp}_2}, 0, 0)$ as a function of γ_1; b line plots of normalised Fisher information for $\gamma_1 = 0°$ and $90°$ as a function of dipole separation. For reference, plots of $K_{02} \pm K_{22}$ are also shown; c as a however a further background intensity count D_b is introduced into the detection process, where $S_0/D_b = 10^2$; d as c albeit including detector scanning in the image plane over a square field of view with spatial extent of $6 \times 0.61\lambda/\text{NA}_2$. The normalisation factor used in d is given by $\int_{\Omega_{\text{im}}} J_\gamma^{\max}(\rho_2) d\rho_2 = \frac{2p_0^2}{\hbar \nu_0} \int_{\Omega_{\text{im}}} K_{01}^{A\,2}(\rho_2) d\rho_2$

Oscillations in the Fisher information as the dipole separation is increased, originally seen for the background case, are still exhibited, albeit the modulation of the oscillations is reduced. The extent of modulation is reduced since the information dips are broadened by the background count hence washing out sharp variations. As D_b is further increased this modulation reduces to a more uniform behaviour.

Scanning the polarimetric detector in the image plane however, also acts to suppress informational oscillations as shown in Fig. 6.14d. Whilst the informational dips at $\gamma_1 = 0°$ and $90°$ are still present, the polarisation variation of the image field helps to overcome informational losses introduced from inference from zero intensities in one or more detectors.

6.5 Physical Constraints in Vectorial Imaging

All electromagnetic fields must satisfy Maxwell's equations. If, as in a noiseless world, infinite measurement accuracy is possible, any set of measured data is consistent with this requirement. In post detection processing however, the presence of noise means any parameters inferred from measurements need not be physically correct. Unbiased estimators will, on average, produce physical parameter estimates, however making such claims for any particular realisation of experimental data is questionable. Furthermore, nonlinear transformations are commonplace in electromagnetic inference, such as the polar matrix decomposition discussed in Sect. 5.4, which may produce biased estimators. Although not necessarily disadvantageous, this does preclude obtaining physical answers.

Incorporating physical constraints into an estimation procedure would be expected to improve the precision of an estimator, since it constitutes a form of a priori knowledge. Furthermore, physical parameter estimates are ensured given any realisation of data. Assessing potential gains by means of the CRLB, it has been shown that inequality constraints, such as a positivity constraint on intensity, give no additional performance gains [18]. Equality constraints, however, do give a reduction in the CRLB (the reduced bound being termed the CCRLB), as was discussed in Sect. 3.4.1. To illustrate the potential improvement, the divergence equation (Eq. (4.1)) will be enforced in a simple $4f$ vectorial imaging system (see Fig. 6.12) in which the object field can be represented by means of the series expansion developed in Sect. 6.2. Although the problem can be formulated in terms of estimation of the object expansion coefficients by a suitable modification to the propagation equations to follow, only estimation of the equivalent expansion coefficients $\mathbf{A}_{N,n}$ in the back focal plane will be considered here for simplicity and brevity.

In free space Eq. (4.1) reduces to $\nabla \cdot \mathbf{E} = 0$. For a single plane wave the divergence equation takes the particularly simple form $\mathbf{s} \cdot \mathbf{E} = 0$ where $\mathbf{s}$ is the unit wavevector defining the direction of the plane wave as discussed in Sect. 4.5. This constraint is thus perhaps easiest to implement over the Gaussian reference sphere in the imaging system, since the electric field vector $\mathbf{e}(\theta_2, \phi_2)$ at each spatial position (θ_2, ϕ_2) can be considered as a Jones vector specifying a single plane wave, such that

$$0 = \sin\theta_2 \cos\phi_2\, e_x(\theta_2, \phi_2) + \sin\theta_2 \sin\phi_2\, e_y(\theta_2, \phi_2) + \cos\theta_2\, e_z(\theta_2, \phi_2). \quad (6.68)$$

Making the substitution $u_2 = \sin\theta_2$ and employing Eq. (6.5) (assuming the object is in-focus such that the m mode index plays no role and is hence dropped) yields the equivalent expression

$$0 = \frac{u_2}{\sqrt{1-u_2^2}} \left[\cos\phi_2\, a_x(u_2, \phi_2) + \sin\phi_2\, a_y(u_2, \phi_2)\right] + a_z(u_2, \phi_2). \quad (6.69)$$

6.5 Physical Constraints in Vectorial Imaging

Applying the expansion $a_j(u_2, \phi_2) = \sum_{N=-\infty}^{\infty}\sum_{n=0}^{\infty} A_{N,n}^j \Phi_{|N|,n}(u_2) \exp(iN\phi_2)$ (as per Eq. (6.6)) and forming the inner product of Eq. (6.69) with $\Phi_{|Q|,q}(u_2)\exp(-iQ\phi_2)$ over the aperture of the detector lens ($NA_2 = \sin\alpha_2 = u_{\alpha_2}$) yields

$$0 = \sum_{N=-\infty}^{\infty}\sum_{n=0}^{\infty} F_{N,n,Q,q}\left[A_{N,n}^x d_{N,Q}^+ - i A_{N,n}^y d_{N,Q}^-\right] + \lambda_{|Q|,q} A_{Q,q}^z, \tag{6.70}$$

where

$$F_{N,n,Q,q} = \int_0^{u_{\alpha_2}} \frac{u_2}{\sqrt{1-u_2^2}} \Phi_{|N|,n}(u_2)\Phi_{|Q|,q}(u_2) u_2 du_2 \tag{6.71}$$

and

$$d_{N,Q}^{\pm} = \frac{\delta_{N,Q+1} \pm \delta_{N,Q-1}}{2}. \tag{6.72}$$

Upon a suitable lexicographical ordering of the mode indices, Eq. (6.70) can be written in matrix form which reads

$$\mathbf{0} = \left(\mathbb{F}^+ \; \mathbb{F}^- \; \mathbb{L}\right)\begin{pmatrix}\mathbf{A}^x \\ \mathbf{A}^y \\ \mathbf{A}^z\end{pmatrix} = \mathbb{G}_\mathbf{A}\mathbf{A}, \tag{6.73}$$

where $\mathbb{F}^{\pm} = [F_{N,n,Q,q} d_{N,q}^{\pm}]$, $\mathbb{L} = \mathrm{diag}[\lambda_{|Q|,q}]$ and $\mathbf{A}^j = [A_{N,n}^j]$. This is precisely of the form discussed in Sect. 3.4 and can hence be used to form a CMLE.

Constrained maximum likelihood estimation of the expansion coefficients $\mathbf{A}$ is however a nonlinear estimation problem, since they must be inferred from intensity measurements. For example, consider a type I scanning microscope, where at each scan point in the image plane a polarimetric measurement is made, using a point detector. Using the eigenfunction expansion, the field at a scan point ρ_k, assuming a coherent system, can be written as a matrix equation viz.

$$\mathbf{E}(\rho_k) = \sum_{N=-\infty}^{\infty}\sum_{n=0}^{\infty} \mathbf{B}_{N,n} \Phi_{|N|,n}\left(\frac{u_{\alpha_2}\rho_k}{\rho_{\max}}\right)\exp(iN\varphi_k), \tag{6.74}$$

$$= \mathbb{P}(\rho_k)\mathbf{B}, \tag{6.75}$$

$$= \mathbb{P}(\rho_k)\mathbb{H}\mathbf{A}. \tag{6.76}$$

Here $\mathbb{H}$ is a system matrix describing the scaling of coefficients, $\mathbf{B}$, in the detector plane to the Gaussian reference sphere, $\mathbf{A}$, as defined by Eq. (6.20) and the elements of $\mathbb{P}(\rho_k)$ give the value of the basis functions $\Phi_{|N|,n}(u_{\alpha_2}\rho_k/\rho_{\max})\exp(iN\varphi_k)$ at each scan point. A polarimetric detection will hence produce a vector of intensity readings given by

$$\mathbf{D}(\rho_k) = \mathbb{T}\,\mathbb{A}\,\text{vec}\left[\mathbf{E}(\rho_k) \otimes \mathbf{E}(\rho_k)^\dagger\right] + \mathbf{n}, \qquad (6.77)$$

$$= \mathbb{T}\,\mathbb{A}\,(\mathbb{P}(\rho_k)^* \otimes \mathbb{P}(\rho_k))(\mathbb{H}^* \otimes \mathbb{H})(\mathbf{A}^* \otimes \mathbf{A}) + \mathbf{n}, \qquad (6.78)$$

where $\mathbb{T}$ is the instrument matrix of the polarimeter, $\mathbb{A}$ is given by Eq. (4.28) and the identities $\text{vec}[\mathbb{XYZ}] = (\mathbb{Z}^T \otimes \mathbb{X})\text{vec}[\mathbb{Y}]$ and $\mathbb{WX} \otimes \mathbb{YZ} = (\mathbb{W} \otimes \mathbb{Y})(\mathbb{X} \otimes \mathbb{Z})$ have been used [7]. $\mathbf{n}$ is a noise vector. Equation (6.78) furthermore identifies a suitable transformation of the problem, from a nonlinear estimation of $\mathbf{A}$ to a linear estimation of $\mathbf{A}^* \otimes \mathbf{A}$, which is much simpler to analyse and has been discussed more fully in Chap. 3.

Since in general the expansion coefficients are complex, then so too are the elements of the parameter vector $\mathbf{A}^* \otimes \mathbf{A}$. As discussed in Sect. 3.2 the parameter vector of interest hence becomes $\mathbf{w} = (\mathbf{A}^* \otimes \mathbf{A}, \mathbf{A} \otimes \mathbf{A}^*)$. It is noted that given a noise free $\mathbf{w}$, it is theoretically possible to deduce all coefficients $\mathbf{A}$ up to a physically irrelevant global phase. The presence of noise again requires some means of statistical estimation, however due to the invariance properties of MLEs, this presents no further complications.

With this reparameterisation of the estimation problem, the constraints expressed by Eq. (6.73) must also be reformulated viz.

$$(\mathbb{G}_\mathbf{A}^* \mathbf{A}^*) \otimes (\mathbb{G}_\mathbf{A}\mathbf{A}) = (\mathbb{G}_\mathbf{A}^* \otimes \mathbb{G}_\mathbf{A})(\mathbf{A}^* \otimes \mathbf{A}) = \mathbf{0} \otimes \mathbf{0}, \qquad (6.79)$$

where the identity $\mathbb{WX} \otimes \mathbb{YZ} = (\mathbb{W} \otimes \mathbb{Y})(\mathbb{X} \otimes \mathbb{Z})$ has again been employed, such that

$$\mathbb{G}_\mathbf{w}\mathbf{w} = \begin{pmatrix} \mathbb{G}_\mathbf{A}^* \otimes \mathbb{G}_\mathbf{A} & \mathbb{O} \\ \mathbb{O} & \mathbb{G}_\mathbf{A} \otimes \mathbb{G}_\mathbf{A}^* \end{pmatrix} \mathbf{w} = \mathbf{0}. \qquad (6.80)$$

Using the preceding formulae the FIM for estimation of $\mathbf{w}$ was calculated in a Gaussian noise regime, assuming noise covariance $\mathbb{K} = \sigma^2\mathbb{I}$. For computational reasons the maximum mode indices were restricted to $n_0 = N_0 = 3$, equating to 63 expansion modes or equivalently 14,112 elements of $\mathbf{w}$.[14] Furthermore, the lenses were assumed to have numerical apertures of 0.95 and 0.0095 respectively equating to a magnification of 100, c was taken to be 10 such that the scan area covered a circular domain with radius of approximately 2.6 Airy units (i.e. $\rho_{\max} \approx 1.59\lambda/\text{NA}_2$), with scan points being separated by 10 μm. From the calculated FIM and the constraint matrix $\mathbb{G}_\mathbf{w}$ the matrices $\mathbb{J}^{-1}$ and $\mathbb{B}$ (see Eq. 3.42) were calculated, representing the CRLB and CCRLB respectively. The trace of these two matrices provide an aggregate measure of the informational limits in the system, in a similar fashion to the Fisher capacity, and are shown as a function of the noise variance σ^2 in Fig. 6.15.

---

[14] The main computational restriction lies in the inversion of the associated 14,112 × 14,112 FIM.

6.5 Physical Constraints in Vectorial Imaging

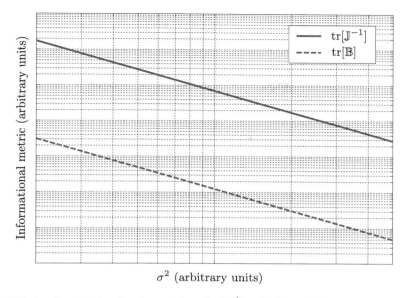

Fig. 6.15 Log-log plot showing the variation of $\text{tr}[\mathbb{J}^{-1}]$ and $\text{tr}[\mathbb{B}]$, which provide an aggregate quantification of the CRLB (*solid*) and CCRLB (*dashed*), with SNR for estimation of 63 generalised prolate spheroidal mode coefficients in the back focal plane of a vectorial imaging system (see text). Units on each axis are arbitrary and scale with the total number of photons present in the system

A logarithmic informational loss as the SNR increases is clearly identifiable, as would be expected, however the principle point of interest is that constrained estimation exhibits a significant potential for improvement over unconstrained estimation, with over two orders of magnitude improvement seen.

6.6 Conclusions

This chapter has considered two alternative approaches to analysing and enhancing the performance of polarisation based imaging systems. Firstly, a new expansion of the electric fields in the focal region of a high NA system in terms of Bessel and generalised circular prolate spheroidal functions was developed. Physically, the generalised prolate spheroidal functions have also been shown to be the eigenfunctions of the focusing operation and hence an inversion formula was also derived and given. This formulation allowed superresolution in imaging systems to be considered from an inverse problem standpoint. The reported method unfortunately proved unsuitable for obtaining this goal however, since although it is theoretically possible to generate an arbitrary distribution in one or two field components, Maxwell's equations dictate this is at the cost of dominant features arising in the unconstrained components.

Synthesis of arbitrary focused fields is however of importance in a number of alternative problems and a number of further inversion examples were presented. The eigenfunction expansion was also demonstrated to possess a number of desirable and physical properties, including maximal energy packing, separability of its component functions in cylindrical coordinates, and fast convergence in the azimuthal, radial and axial directions. Although it is not impossible to solve for a general plane in the focal region, a restriction to the focal plane does greatly simplify the calculations. A number of further caveats to the use of the inversion formula have also been presented, including consideration of the degrees of freedom present in the system which restricts how fully the focal distribution can be specified to maintain physicality.

Perhaps the most important result to come from the eigenfunction representation stems from the considerations of the energy contained within the field specification area. By considering the meaning of the prolate functions, a means to construct "reasonable" focal distributions becomes apparent. In the sense of energy distribution, an optimum field can be constructed by using only the prolate functions with large eigenvalues as a basis in the focal plane. This also produces better inversion results due to reduced noise amplification.

Although much focus has been placed on "exact" inversion, the developed formulae are also highly suitable for numerical optimisation, for which there already exists a vast range of tools and knowledge. Such suitability arises since only relatively few orders are required for reasonably accurate results and hence the number of optimisation parameters i.e. expansion coefficients, is also small. This is especially true for synthesis of axial or circularly symmetric transverse distributions since in this case only the $N = 0$ modes have non-zero coefficients.

Due to the difficulty in obtaining high fidelity between an image and the object under study, a second, Fisher information based, approach was also developed, which quite naturally highlights the greater potential in imaging as opposed to wide area integrated measurements. In this vein the performance of a simple polarisation microscope imaging a single electric dipole was considered and a number of polarisation measurement architectures evaluated. Whilst it was found that when attempting to determine the orientation of a single dipole a background free, crossed polariser configuration performs optimally, informational losses at particular (although highly predictable) orientations were seen to occur when additional noise sources are introduced. Information losses of this nature were also seen in more complex architectures.

Introduction of a second, extraneous dipole in the object plane, can, if incoherently radiating with respect to the first, act as such a background source, thus having a detrimental effect on the measurement process. Although similar effects can be seen for a second coherently radiating dipole, the extent of performance degradation becomes dependent on the relative configuration of the two dipoles producing a more complex behaviour. In essence however, all informational losses are seen to occur when inference is based on null readings, which are highly susceptible to any additional perturbations. Perhaps by way of extolling the virtues and power of imaging systems, if scanning measurements are taken and estimation based on the subsequent image formed, a reduction in the complexity of the dipole crosstalk attributes is seen, partly due to the averaging introduced in such measurements, but also due to the greater redundancy introduced.

6.6 Conclusions

Image reconstruction or parameter estimation from electromagnetic imaging measurements, is however subject to Maxwell's equations; a factor which is commonly overlooked in reconstruction algorithms. As such consideration was finally given to the potential improvements achievable were such constraints introduced into the associated algorithms. Although computational expensive, significant possible gains were demonstrated by comparison of metrics derived from the CRLB and the CCRLB. Further avenues of research however still lie open in this vein, not only in the design and practical implementation of suitable routines, but also in imposition of further constraints and a priori knowledge.

References

1. G.P. Agrawal, D.N. Pattanayak, Gaussian beam propagation beyond the paraxial approximation. J. Opt. Soc. Am. **69**, 575–578 (1979)
2. G. Arfken, *Mathematical Methods for Physicists,* 3rd edn. (Elsevier Academic Press, New York, 1985)
3. Z. Bomzon, G. Biener, V. Kleiner, E. Hasman, Radially and azimuthally polarized beams generated by space-variant dielectric subwavelength gratings. Opt. Lett. **27**, 285–287 (2002)
4. M. Born, E. Wolf, *Principles of Optics,* 7th edn. (Cambridge University Press, Cambridge, 1980)
5. J.J.M. Braat, P. Dirksen, A.J.E.M. Janssen, A.S. van de Nes, Extended Nijboer-Zernike representation of the vector field in the focal region of an aberrated high-aperture optical system. J. Opt. Soc. Am. A **20**, 2281–2292 (2003)
6. U. Brand, G. Hester, J. Grochmalicki, R. Pike, Super-resolution in optical data storage. J. Opt. A Pure Appl. Opt. **1**, 794–800 (1999)
7. M. Brookes, The matrix reference manual (2005). http://www.ee.ic.ac.uk/hp/staff/dmb/matrix/intro.html.
8. D.R. Chowdhury, K. Bhattacharya, S. Sanyal, A.K. Chakraborty, Performance of a polarization-masked lens aperture in the presence of spherical aberration. J. Opt. A Pure Appl. Opt. **4**, 98–104 (2002)
9. T. di Francia, Super-gain antennas and optical resolving power. Nuovo Cimento **9**, 426–438 (1952)
10. E.R. Dowski Jr., W.T. Cathey, Extended depth of field through wave-front coding. Appl. Opt. **34**, 1859–1866 (1995)
11. M. Endo, Pattern formation method and exposure system. Patent 7,094,521, 2006
12. P.E. Falloon, Hybrid computation of the spheroidal harmonics and application to the generalized hydrogen molecular ion problem, Ph.D. thesis, 2001
13. P.B. Fellgett, E.H. Linfoot, On the assessment of optical images. Philos. Trans. R. Soc. Lond. A **247**, 369–407 (1955)
14. M.R. Foreman, S.S. Sherif, P.R.T. Munro, P. Török, Inversion of the Debye-Wolf diffraction integral using an eigenfunction representation of the electric fields in the focal region. Opt. Express **16**, 4901–4917 (2008)
15. B.R. Frieden, *Evaluation, Design and Extrapolation Methods for Optical Signals, Based on the Prolate Functions.* Progress in Optics IX (North-Holland Publishing Co., Amsterdam, 1971)
16. B.R. Frieden, *Physics from Fisher Information: A Unification* (Cambridge University Press, Cambridge, 1998)
17. D. Gabor, A new microscopic principle. Nature (London) **161**, 777–778 (1948)
18. J.D. Gorman, A.O. Hero, Lower bounds for parametric estimation with constraints. IEEE Trans. Inform. Theory **26**, 1285–1301 (1990)

19. I.S. Gradshteyn, I.M. Ryzhik, *Table of Integrals, Series and Products* (Elsevier Academic Press, New York, 1980)
20. T. Ha, T. Enderle, D.S. Chemla, P.R. Selvin, S. Weiss, Single molecule dynamics studied by polarization modulation. Phys. Rev. Lett. **77**, 3979–3982 (1996)
21. P.C. Hansen, *Rank-Deficient and Discrete Ill-Posed Problems: Numerical Aspects of Linear Inversion* (SIAM, Philadelphia, 1997)
22. Z.S. Hegedus, V. Sarafis, Superresolving filters in confocally scanned imaging systems. J. Opt. Soc. Am. A **3**, 1892–1896 (1986)
23. J.C. Heurtley, Hyperspheroidal functions-optical resonators with circular mirrors, in *Proceedings of Symposium on Quasi-Optics*, ed. by J. Fox (Polytechnic Press, New York, 1964), pp. 367–371
24. P.D. Higdon, P. Török, T. Wilson, Imaging properties of high aperture multiphoton fluorescence scanning microscopes. J. Micros. **193**, 127–141 (1999)
25. S. Inoué, Exploring Living Cells and Molecular Dynamics with Polarized Light Microscopy, 1st edn. In Török and Kao [57], 2007, ch. 1.
26. R. Kant, An analytical solution of vector diffraction for focusing optical systems with Seidel aberrations. J. Mod. Opt. **40**, 2293–2310 (1993)
27. B. Karczewski, E. Wolf, Comparison of three theories of electromagnetic diffraction at an aperture. J. Opt. Soc. Am **56**, 1207–1219 (1966)
28. S. Kirkpatrick, C.D. Gelatt, M.P. Vecchi, Optimization by simulated annealing. Science **220**, 671–680 (1983)
29. D. Lara, C. Paterson, Stokes polarimeter optimization in the presence of shot and Gaussian noise. Opt. Express **17**, 21240–21249 (2009)
30. W. Latham, M. Tilton, Calculation of prolate functions for optical analysis. Appl. Opt. **26**, 2653–2658 (1987)
31. W.H. Lee, *Computer-Generated Holograms: Techniques and Applications,* Progress in Optics XVI (North-Holland Publishing Co., Amsterdam, 1978)
32. C.W. McCutcheon, Generalised aperture and the three-dimensional diffraction image. J. Opt. Soc. Am. **54**, 240–244 (1964)
33. P.R.T. Munro, Application of numerical methods to high numerical aperture imaging. Ph.D. thesis, 2006
34. Y. Mushiake, K. Matsumura, N. Nakajima, Generation of radially polarized optical beam mode by laser oscillation. Proc. IEEE **60**, 1107–1109 (1972)
35. M.A.A. Neil, F. Massoumian, R. Juškaitis, T. Wilson, Method for the generation of arbitrary complex vector wave fronts. Opt. Lett. **27**, 1929–1931 (1990)
36. M.A.A. Neil, T. Wilson, R. Juškaitis, A wavefront generator for complex pupil function synthesis and point spread function engineering. J. Micros. **197**, 219–223 (2000)
37. R.J. Ober, S. Ram, E.S. Ward, Localization accuracy in single-molecule microscopy. Biophys. J. **86**(2), 1185–1200 (2004)
38. J. Ojeda-Castañeda, L.R. Berriel-Valdos, E. Montes, Spatial filter for increasing the depth of focus. Opt. Lett. **10**, 520–522 (1987)
39. R. Oldenbourg, A new view on polarization microscopy. Nature **381**, 811–812 (1996)
40. R. Oldenbourg, G. Mei, New polarized light microscope with precision universal compensator. J. Micros. **180**, 140–147 (1995)
41. R. Oldenbourg, P. Török, Point-spread functions of a polarizing microscope equipped with high-numerical-aperture lenses. Appl. Opt. **39**, 6325–6331 (2000)
42. R. Pike, D. Chana, P. Neocleous, S. Jiang, Superresolution in scanning optical systems, 1st edn. In Török and Kao [57], 2007, Ch. 4.
43. T.C. Poon, M. Motamedi, Optical/digital incoherent image processing for extended depth of field. Appl. Opt. **26**, 4612–4615 (1987)
44. S. Ram, E.S. Ward, R.J. Ober, Beyond Rayleigh's criterion: a resolution measure with application to single-molecule microscopy. Proc. Natl Acad. Sci. USA **103**, 4457–4462 (2006)
45. B. Richards, E. Wolf, Electromagnetic diffraction in optical systems. II. Structure of the image field in an aplanatic system. Trans. Opt. Inst. Pet. **253**, 358–379 (1959)

46. A. Rohrbach, J. Huisken, E.H.K. Stelzer, Optical trapping of small particles, 1st edn. In Török and Kao [57], 2007, ch. 15.
47. C.J.R. Sheppard, A. Choudhury, Image formation in the scanning microscope. J. Mod. Opt. **24**(10), 1051–1073 (1976)
48. C.J.R. Sheppard, P. Török, Efficient calculation of electromagnetic diffraction in optical systems using a multipole expansion. J. Mod. Opt. **44**, 803–818 (1997)
49. S.S. Sherif, W.T. Cathey, Depth of field control in incoherent hybrid imaging systems, 1st edn. In Török and Kao [57], 2007, ch. 5.
50. S.S. Sherif, M.R. Foreman, P. Török, Eigenfunction expansion of the electric fields in the focal region of a high numerical aperture focusing system. Opt. Express **16**, 3397–3407 (2008)
51. S.S. Sherif, P. Török, Pupil plane masks for super-resolution in high-numerical-aperture focusing. J. Mod. Opt. **51**, 2007–2019 (2004)
52. S.S. Sherif, P. Török, Eigenfunction representation of the integrals of the Debye-Wolf diffraction formula. J. Mod. Opt. **52**, 857–876 (2005)
53. B. Sick, B. Hecht, L. Novotny, Orientational imaging of single molecules by annular illumination. Phys. Rev. Lett. **85**, 4482–4485 (2000)
54. D. Slepian, Prolate spheroidal wave functions, Fourier analysis and uncertainty IV Extensions to many dimensions; generalised prolate spheroidal functions. Bell System Tech. J. **43**, 3009–3057 (1964)
55. S.C. Tidwell, D.H. Ford, W.D. Kimura, Generating radially polarized beams interferometrically. Appl. Opt. **29**, 2234–2239 (1990)
56. P. Török, P.D. Higdon, T. Wilson, Theory for confocal and conventional microscopes imaging small dielectric scatterers. Opt. Commun. **45**, 1681–1698 (1998)
57. P. Török, F.-J. Kao (eds.), *Optical Imaging and Microscopy—Techniques and Advanced Systems*, 1st edn. (Springer, New York, 2007)
58. P. Török, P. Varga, Electromagnetic diffraction of light focused through a stratified medium. Appl. Opt. **36**, 2305–2312 (1997)
59. K.C. Toussaint Jr., S. Park, J.E. Jureller, N.F. Scherer, Generation of optical vector beams with a diffractive optical element interferometer. Opt. Lett. **30**, 2846–2848 (2005)
60. H.C. van de Hulst, *Light Scattering by Small Particles* (Dover Publications, Dover, 1981)
61. A.S. van de Nes, Rigorous electromagnetic field calculations for advanced optical systems. Ph.D. thesis, 2005
62. G.N. Watson, *A Treatise on the Theory of Bessel Functions* (Cambridge University Press, Cambridge, 1995)
63. W.T. Welford, Use of annular apertures to increase focal depth. J. Opt. Soc. Am. **50**, 749–753 (1960)
64. W.T. Welford, On the relationship between the modes of image formation in scanning microscopy and conventional microscopy. J. Micros. **96**, 105–107 (1972)
65. T. Wilson, C. Sheppard, *Theory and Practice of Scanning Optical Microscopy* (Academic Press, London, 1984)
66. E. Wolf, Electromagnetic diffraction in optical systems I An integral representation of the image field. Proc. R. Soc. Lond. A **253**, 349–357 (1959)
67. K.S. Youngworth, T.G. Brown, Focusing of high numerical aperture cylindrical vector beams. Opt. Express **7**, 77–87 (2000)
68. S.-S. Yu, B.J. Lin, A. Yen, C.-M. Ke, J. Huang, B.-C. Ho, C.-K. Chen, T.-S. Gau, H.-C. Hsieh, Y.-C. Ku, Thin-film optimization strategy in high numerical aperture optical lithography I—principles. J. Microlith. Microfab. Microsyst. **4**, 043003 (2005)
69. T. Zolezzi, *Well-posedness Criteria in Optimization with Application to the Calculus of Variations*. Nonlinear Analysis: Theory, Methods and Applications, vol. 25 (Elsevier Science Ltd., Oxford, 1995), pp. 437–453

Chapter 7
Multiplexed Optical Data Storage (MODS)

Not all bits have equal value.
Carl E. Sagan

Compact discs (CDs), offering a total of 640 MB of storage, revolutionised the information storage industry when first introduced in 1982. The CD has since proliferated worldwide and become a cornerstone of optical data storage (ODS). Although still widely used, the storage capacity of CDs proved insufficient for a number of consumer needs. Growing storage demands have hence driven the industry to develop the digital versatile disc (DVD) in 1995 and, more recently, the high density DVD (HD-DVD) and Blu-ray disc (BD) in 2006 (see [10] for fuller details). Whilst alternative storage media exist capable of meeting market demands, such as magnetic drives or solid state random access memory (RAM), ODS systems and media are inexpensive to produce and master ($\sim$ £0.05 for a DVD disc) and therefore continue to be of major significance.

Advances in ODS technology have been realised by successive increases in the NA of the illumination optics (0.45, 0.6, and 0.85 for CD, DVD and BD respectively) and decreases in the wavelength of light employed (780, 650 and 405 nm for CD, DVD and BD respectively)[1] so as to reduce the size of the diffraction limited focal spot on the disc. Decreasing the size of the focal spot in turn allows data pits to be reduced in size with a resulting increase in information density and hence total capacity of an optical disc.

Potential improvements in ODS via NA increase and wavelength reduction are, however, now reaching their limits. Use of shorter wavelengths is currently unfeasible since common optical media used in lenses, such as BK7, exhibit poor transmission at wavelengths below approximately 330 nm. Further increases in NA are also problematic since the maximum NA achievable for a lens in air is unity. Achieving larger NAs requires use of immersion lenses, which couple evanescent waves scattered from a data pit into the readout optics, hence circumventing the diffraction limit [18]. Some

---

[1] Improved encoding algorithms have also led to increased data capacities, but these however fall outside the remit of this thesis and will not be considered.

success has already been achieved using a solid immersion lens (SIL), for example a 50 GB storage capacity has been demonstrated under laboratory conditions [31], however use of immersion lenses still poses significant engineering problems. Difficulties arise, for instance, since a SIL based system requires maintaining an air gap of a few tens of nanometers between the SIL and the disc surface, whilst the disc rotates at high speed. Robustness of SIL systems outside the laboratory must thus first be improved before commercial exploitation is possible.

For improvements beyond those achievable using immersion lenses, recourse must be made to alternative strategies, such as multiplexing in which a single pit can store more than a single bit of information. Accordingly a research consortium was tasked in 2001 with identifying the most viable multiplexing technique, including evaluation of techniques based on amplitude, phase and orbital angular momentum encoding [27]. Key to the findings of this consortium was the conclusion that polarisation multiplexing presented the most viable approach.

Utilising the polarisation state of light to multiplex information on an optical disc requires the creation of scattering structures with a form birefringence. Form birefringence typically arises from an asymmetry in a structure. Rectangular pits were, for example, considered in [17] whilst rod-like particles were considered in [6, 24], which have all demonstrated a dependence of the scattered field on the incident polarisation. Numerical simulations have furthermore demonstrated the importance of the polarisation of the illuminating field [3, 4, 9, 11, 15, 19] in determining the readout properties of an ODS system using a variety of scattering structures, including grooves, pits and bumps. None have, however, considered the exploitation of polarisation as an information carrying channel.

Within this context this chapter considers a simplified electromagnetic model of light scattering. In particular the rectangular pits proposed in [17] and shown in Fig. 7.1a are considered since the experimental results presented therein have shown a potential sevenfold increase in data capacity over existing BD technology and thus hold significant promise. The scattering model is set up with the aim of determining the key principles by which a polarisation multiplexed system (henceforth referred to simply as a MODS system) operates and how such principles can be fully utilised. As such the scattering calculations are finally used as a basis for numerical optimisation of the geometry of the scattering structure.

7.1 Electromagnetic Scattering Calculations

Modelling and characterisation of the proposed MODS system is imperative for successful implementation. As the need for storage capabilities has increased the size of the associated data pits has decreased to sub-wavelength dimensions requiring more sophisticated models to obtain accurate results. Predominantly these models employ rigorous coupled wave theory [20, 21], the finite difference time domain (FDTD) method [14], the finite element method (FEM) [7, 29] and the boundary integral method (BIM) which all derive from Maxwell's equations.

7.1 Electromagnetic Scattering Calculations

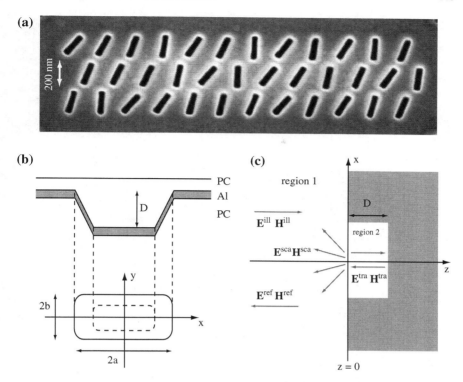

Fig. 7.1 **a** Scanning electron microscope image of elongated data pits as proposed for a polarisation multiplexed optical data storage system. Pits measure 200 × 50 nm. **b** Simple schematic of a data pit on the surface of an optical disc. **c** Geometry of mode matching calculations at the surface of a MODS disc

Rigorous modelling techniques, such as these, however require significant computational resources for a single calculation and are thus ill-suited for the optimisation calculations to be performed in Sect. 7.3. Instead a modal method in which the field in different regions of space are decomposed into a basis of electromagnetic field distributions, or modes, which satisfy Maxwell's equations is used. Modal methods have been proposed for various diffraction and scattering problems, for example [2, 3, 12, 23], since they are numerically more tractable than exact techniques, such as FDTD, and because they reduce the full 3D problem to a 2D equivalent. The mode expansion presented here closely follows that presented in [2]. Furthermore, simulation results shown were calculated using code kindly provided by J. Brok.

7.1.1 System Description

Before numerical simulations can be undertaken it is first necessary to specify the geometry of the scattering problem. Data pits on optical discs are commonly formed on a polycarbonate substrate, with an aluminium coating with profile as shown in Fig. 7.1b. A protective polycarbonate coating is then applied. Simulations presented in this chapter, however adopt a simplified model of a single MODS data pit. Specifically a $2a \times 2b$ rectangular pit opening is used, whilst the sides of the pit are assumed to be perpendicular to the disc surface and to extend to a depth of D. The aluminium layer is modelled as a perfect conductor and the plane of the disc surface is considered to lie perpendicular to the optical axis at the focal plane of a high numerical aperture lens. Although the protective polycarbonate layer is also neglected for simplicity, this could potentially be incorporated into the system by using the focusing formalism described in [28]. With a view to backward compatibility with BD technology, a wavelength of $\lambda = 405$ nm is used in all simulations.

7.1.2 Mode Expansion Theory

The mode expansion detailed in [2, 3] distinguishes between two regions in space, namely the region above the disc interface (designated region 1) and the region within the pit volume (region 2), as shown in Fig. 7.1c. Regions 1 and 2 are assumed to both possess a refractive index of unity (hence $\epsilon_1 = \epsilon_2 = \epsilon_0$) and to be non-magnetic ($\mu_1 = \mu_2 = \mu_0$). Consideration need not be given to the field within the conducting material of the disc surface, since the assumption of perfect conductivity automatically ensures the fields are identically zero.

In region 1 three components of the total field, $\mathbf{E}_1(\mathbf{r} = x, y, z)$, can be identified: the incident illumination field, $\mathbf{E}^{ill}(\mathbf{r})$; the field reflected from the surface of the disc if the pit were not present, $\mathbf{E}^{ref}(\mathbf{r})$; and finally the difference field arising from scattering from the data pit, $\mathbf{E}^{sca}(\mathbf{r})$. Hence

$$\mathbf{E}_1(\mathbf{r}) = \mathbf{E}^{ill}(\mathbf{r}) + \mathbf{E}^{ref}(\mathbf{r}) + \mathbf{E}^{sca}(\mathbf{r}) \tag{7.1}$$

and similarly for the magnetic field

$$\mathbf{H}_1(\mathbf{r}) = \mathbf{H}^{ill}(\mathbf{r}) + \mathbf{H}^{ref}(\mathbf{r}) + \mathbf{H}^{sca}(\mathbf{r}) . \tag{7.2}$$

Within region 2 there will exist a transmitted electric field $\mathbf{E}^{tra}(\mathbf{r})$ and an associated transmitted magnetic field $\mathbf{H}^{tra}(\mathbf{r})$, which will itself be composed of forward travelling and backward travelling components[2] arising from reflections from the back surface of the pit.[3]

As dictated by the appropriate boundary conditions in each region the various contributions to the total field are expressed as a superposition of basis functions,

---

[2] Evanescent modes that may exist in the pit volume, although not strictly propagating, will be composed of waves which decay in the positive and negative z direction hence allowing analogous arguments to be applied.

[3] Possible reflections from the sides of the pit are neglected for simplicity.

7.1 Electromagnetic Scattering Calculations

which themselves form a complete[4] orthogonal set. These expansions are more fully introduced in the next two sections.

7.1.2.1 Mode Expansion Above the Interface

Considering first the field decomposition above the disc surface, it is noted that propagating (and evanescent) plane waves of the form $\mathbf{E} = \mathbf{E}_0 \exp(\pm i \mathbf{k} \cdot \mathbf{r})$ satisfy Maxwell's equations. An angular spectrum representation of the field, as discussed in Sects. 4.2.2 and 4.5, is thus suitable such that

$$\mathbf{E}^\eta(\mathbf{r}) = \int_{-\infty}^{\infty} \int_{-\infty}^{\infty} \mathbf{E}_0^\eta(\mathbf{k}) \exp(\pm i \mathbf{k} \cdot \mathbf{r}) dk_x dk_y , \qquad (7.3)$$

where the superscript $\eta =$ ill, ref, sca denotes the illumination, reflected and scattered field components respectively. The appropriate choice of sign in the exponent is dependent on η. The illuminating field must, by definition, be formed from only forward propagating waves (i.e. the positive sign adopted), whilst the specularly reflected field is therefore automatically described by outward propagating plane waves (a negative exponent). Sommerfeld's radiation condition, which states that no energy can be radiated from infinity, (or conversely that energy scattered from an object must be radiated to infinity) [25], then finally implies that the scattered field is comprised of outward propagating waves only, that is the negative exponent is again adopted.

There are however two modifications to this integral which will prove useful in later calculations. Firstly, by taking heed of the geometry of the scattering problem for which the mode expansion will be ultimately used, it proves advantageous to decompose a single plane wave into its s and p polarised components,[5] defined respectively as the field for which the electric field vector lies perpendicular and parallel to the plane containing the corresponding ray and the optical axis. Equivalently, s and p polarised plane waves can be defined as those having a zero longitudinal electric and magnetic field component respectively.

Secondly, under the assumption that the focal length of the illuminating and collecting lens in the disc illumination and readout system is large in comparison to the wavelength of the light used, any evanescent waves that may arise in region 1 can be neglected as their amplitude will be heavily attenuated giving a null signal on the readout detector. Hence it is possible to restrict the domain of integration of Eq. (7.3) to consider only propagating plane waves. If a SIL system were being used this assumption would have to be discarded and the evanescent wave coupling also considered. Furthermore, the finite numerical aperture of the focusing (and collector)

---

[4] Complete in the sense that an arbitrary field satisfying Maxwell's equations and the boundary conditions can be represented as an appropriate superposition of the modes. This does not necessarily imply completeness in the strict mathematical sense.

[5] Given an arbitrary field vector it is possible to calculate the s and p components using the rotation matrix introduced in Sect. 4.3.

lens implies that only a finite angular range of rays are collected. Specifically only rays for which $k_x^2 + k_y^2 \leq k^2 \text{NA}^2$ are present in the illuminating (and hence also the specularly reflected) beam, where NA is the numerical aperture of the focusing lens (see Sect. 4.5), hence allowing the integration domain to be further limited. Although plane waves propagating at large angles to the optical axis may exist in the scattered field, the collector lens (assumed to be the same as the focusing lens) will also not collect such components, implying their contribution to the readout signal can be safely ignored.

With these considerations in mind Eq. (7.3) becomes

$$\mathbf{E}^\eta(\mathbf{r}) = \sum_{\nu=s,p} \iint_{k_x^2+k_y^2 \leq k^2 \text{NA}^2} a^{\eta,\nu}(\mathbf{k}) \mathbf{E}^\nu(\mathbf{k}) \, dk_x dk_y \,, \tag{7.4}$$

where ν denotes the s and p polarised components (see Appendix B),

$$\mathbf{E}^s(\mathbf{k}) = \begin{cases} \dfrac{\omega\mu}{2\pi\sqrt{k_x^2+k_y^2}} \begin{pmatrix} k_y \\ -k_x \\ 0 \end{pmatrix} & \text{for } k_x^2 + k_y^2 > 0 \,, \\ \dfrac{\omega\mu}{2\pi} \begin{pmatrix} 0 \\ -1 \\ 0 \end{pmatrix} & \text{for } k_x^2 + k_y^2 = 0 \,, \end{cases} \tag{7.5}$$

$$\mathbf{E}^p(\mathbf{k}) = \begin{cases} \dfrac{k_z}{2\pi\sqrt{k_x^2+k_y^2}} \sqrt{\dfrac{\mu}{\epsilon}} \begin{pmatrix} k_x \\ k_y \\ -(k_x^2+k_y^2)/k_z \end{pmatrix} & \text{for } k_x^2 + k_y^2 > 0 \,, \\ \dfrac{k}{2\pi} \sqrt{\dfrac{\mu}{\epsilon}} \begin{pmatrix} 1 \\ 0 \\ 0 \end{pmatrix} & \text{for } k_x^2 + k_y^2 = 0 \,. \end{cases} \tag{7.6}$$

and $a^{\eta,\nu}(\mathbf{k})$ denotes the various scalar angular spectra.

7.1.2.2 Mode Expansion Below the Interface

When considering the physically supported field distributions in region 2, Maxwell's equations must be solved allowing for the presence of the perfectly conducting material bounding the region. Analytic solutions exist, which shall henceforth be referred to as waveguide modes, a derivation of which is given in Appendix B. Here, however only the final form of the modes are given for ease of reference.

Waveguide modes, in a manner similar to that discussed for plane waves, can be categorised as either transverse electric (TE) or transverse magnetic (TM) for which $E_z^{\text{TE}}(\mathbf{r}) = 0$ and $H_z^{\text{TM}}(\mathbf{r}) = 0$, for all $\mathbf{r}$ respectively. In particular, for a pit oriented

7.1 Electromagnetic Scattering Calculations

as shown in Fig. 7.1b, the TM modes are of the form

$$E^{TM}_{xmn}(\mathbf{r}) = \frac{2i\kappa_z}{k^2 - \kappa_z^2} \sin[\kappa_z(z-D)] \frac{\partial}{\partial x} X^{TM}_m(x) Y^{TM}_n(y), \quad (7.7a)$$

$$E^{TM}_{ymn}(\mathbf{r}) = \frac{2i\kappa_z}{k^2 - \kappa_z^2} \sin[\kappa_z(z-D)] X^{TM}_m(x) \frac{\partial}{\partial y} Y^{TM}_n(y), \quad (7.7b)$$

$$H^{TM}_{xmn}(\mathbf{r}) = \frac{-2i\epsilon\omega}{k^2 - \kappa_z^2} \cos[\kappa_z(z-D)] X^{TM}_m(x) \frac{\partial}{\partial y} Y^{TM}_n(y), \quad (7.7c)$$

$$H^{TM}_{ymn}(\mathbf{r}) = \frac{2i\epsilon\omega}{k^2 - \kappa_z^2} \cos[\kappa_z(z-D)] \frac{\partial}{\partial x} X^{TM}_m(x) Y^{TM}_n(y), \quad (7.7d)$$

$$E^{TM}_{zmn}(\mathbf{r}) = 2\cos[\kappa_z(z-D)] X^{TM}_m(x) Y^{TM}_n(y), \quad (7.7e)$$

for $-a \leq x \leq a$, $-b \leq y \leq b$, $0 \leq z \leq D$, whilst the equivalent expressions for TE modes are

$$E^{TE}_{xmn}(\mathbf{r}) = \frac{-2i\mu\omega}{k^2 - \kappa_z^2} \sin[\kappa_z(z-D)] X^{TE}_m(x) \frac{\partial}{\partial y} Y^{TE}_n(y), \quad (7.8a)$$

$$E^{TE}_{ymn}(\mathbf{r}) = \frac{2i\mu\omega}{k^2 - \kappa_z^2} \sin[\kappa_z(z-D)] \frac{\partial}{\partial x} X^{TE}_m(x) Y^{TE}_n(y), \quad (7.8b)$$

$$H^{TE}_{xmn}(\mathbf{r}) = \frac{2i\kappa_z}{k^2 - \kappa_z^2} \cos[\kappa_z(z-D)] \frac{\partial}{\partial x} X^{TE}_m(x) Y^{TE}_n(y), \quad (7.8c)$$

$$H^{TE}_{ymn}(\mathbf{r}) = \frac{2i\kappa_z}{k^2 - \kappa_z^2} \cos[\kappa_z(z-D)] X^{TE}_m(x) \frac{\partial}{\partial y} Y^{TE}_n(y), \quad (7.8d)$$

$$H^{TE}_{zmn}(\mathbf{r}) = 2\sin[\kappa_z(z-D)] X^{TE}_m(x) Y^{TE}_n(y), \quad (7.8e)$$

where

$$X^{\mu}_m(x) = [1 \pm (-1)^m] i \sin(\kappa_x x) + [1 \mp (-1)^m] \cos(\kappa_x x), \quad (7.9)$$

$$Y^{\mu}_n(y) = [1 \pm (-1)^n] i \sin(\kappa_y y) + [1 \mp (-1)^n] \cos(\kappa_y y), \quad (7.10)$$

for $\mu =$ TM and TE modes respectively, $(\kappa_x, \kappa_y) = (m\pi/2a, n\pi/2b)$ for integer m and n, and $\kappa_z = \sqrt{k^2 - \kappa_x^2 - \kappa_y^2}$. A general field in region 2 can therefore be represented as an expansion of waveguide modes, viz.

$$\mathbf{E}(\mathbf{r}) = \sum_{\mu} \sum_{m,n} b^{\mu}_{mn} \mathbf{E}^{\mu}_{mn}(\mathbf{r}), \quad (7.11a)$$

$$\mathbf{H}(\mathbf{r}) = \sum_{\mu} \sum_{m,n} b^{\mu}_{mn} \mathbf{H}^{\mu}_{mn}(\mathbf{r}). \quad (7.11b)$$

The lowest TE mode which does not give a zero field is the $(m, n) = (0, 1)$ or $(1, 0)$ mode (denoted TE_{01} or TE_{10}), whilst the lowest non-zero TM mode is the $(1, 1)$ mode.

The form of these waveguide modes differ slightly from those given in [2] due to a shift in the origin of the coordinate system to the center of the data pit. This shift has been introduced because it is more consistent with the coordinate system used in the Debye-Wolf integral (Eq. (4.46)) used to calculate the focused field incident upon the data pit. It should be noted that Eqs. (7.7) and (7.8) are only valid in region 2, but are however also valid for data pits whose long axis is not parallel to the x-axis, if considered in a coordinate system rotated in the same manner as the pit.

It is insightful to briefly consider the nature of the waveguide modes for waveguides of sub-wavelength dimensions. Solutions to the wave equation are assumed to have the form $X(x) Y(y) \exp(\pm i \kappa_z z)$ (see Appendix B), where κ_z is the propagation constant for a particular mode. Propagating modes therefore correspond to real κ_z, whilst evanescent modes correspond to imaginary κ_z, i.e.

$$\kappa_z^2 \geq 0 \iff k^2 \geq \kappa_x^2 + \kappa_y^2 \iff \text{propagating modes,}$$
$$\kappa_z^2 < 0 \iff k^2 < \kappa_x^2 + \kappa_y^2 \iff \text{evanescent modes.}$$

If $\kappa_z = 0$ the mode is said to be at cutoff. Recalling that for monochromatic light

$$\kappa_z^2 = k^2 - \left(\frac{m\pi}{2a}\right)^2 - \left(\frac{n\pi}{2b}\right)^2, \tag{7.12}$$

higher order modes (i.e. larger m and n) correspond to smaller propagation constants, however if the pit dimensions are small enough even the lowest order mode (either TE_{01} or TE_{10} if $a < b$ or $a > b$ respectively) will be evanescent. Formally it can be shown from Eq. (7.12), that if the largest dimension of the waveguide is less than half a wavelength (in the core medium) *all* waveguide modes will be evanescent in nature. Adopting the values $a = 100$ nm and $b = 25$ nm as used in [17] it is evident that *these conditions are fulfilled for the proposed MODS data pit*, a fact that will be seen to have particular consequences for designing the pit geometry as discussed in Sect. 7.3.

7.1.2.3 Mode Matching at the Interface

Having explicitly given mode expansions for the fields above the disc surface and within the data pit, it is now necessary to consider the coupling of the modes between the two regions. Again Maxwell's equations and the resulting boundary conditions provide a means to investigate this.

7.1 Electromagnetic Scattering Calculations

At the boundary between the two regions the fields must match meaning

$$\mathbf{E}^{\text{tra}}(x, y, 0) = \mathbf{E}^{\text{ill}}(x, y, 0) + \mathbf{E}^{\text{ref}}(x, y, 0) + \mathbf{E}^{\text{sca}}(x, y, 0), \quad (7.13a)$$
$$\mathbf{H}^{\text{tra}}(x, y, 0) = \mathbf{H}^{\text{ill}}(x, y, 0) + \mathbf{H}^{\text{ref}}(x, y, 0) + \mathbf{H}^{\text{sca}}(x, y, 0). \quad (7.13b)$$

The reflected field $\mathbf{E}^{\text{ref}}$ is, however, defined as that present in the absence of a data pit, i.e. for reflection from a perfect planar conductor. Applying the boundary conditions for a perfect conductor (c.f. Eqs. (B.25) and (B.26)) then gives

$$\mathbf{E}^{\text{ill}}_{\parallel}(x, y, 0) + \mathbf{E}^{\text{ref}}_{\parallel}(x, y, 0) = 0, \quad (7.14)$$

where $\parallel$ denotes the field components that are tangential to the conductor surface. The boundary conditions for the data pit mode matching problem can thus be written

$$\mathbf{E}^{\text{tra}}_{\parallel}(x, y, 0) = \mathbf{E}^{\text{sca}}_{\parallel}(x, y, 0), \quad (7.15a)$$
$$\mathbf{H}^{\text{tra}}_{\parallel}(x, y, 0) = \mathbf{H}^{\text{ill}}_{\parallel}(x, y, 0) + \mathbf{H}^{\text{ref}}_{\parallel}(x, y, 0) + \mathbf{H}^{\text{sca}}_{\parallel}(x, y, 0). \quad (7.15b)$$

For a given illumination, the fields $\mathbf{H}^{\text{ill}}_{\parallel}(x, y, 0)$ and $\mathbf{H}^{\text{ref}}_{\parallel}(x, y, 0)$ can be calculated and are thus assumed known. Equations (7.15) represent two vector equations, with two sets of unknowns, namely the angular spectrum $a^{\text{sca},\nu}$ and the expansion coefficients b^{μ}_{mn}. To solve this system of simultaneous equations the orthogonality properties of plane waves can be used as follows.

Consider first forming the inner product of Eq. (7.15a) with an arbitrary plane wave, viz.

$$\langle \mathbf{E}^{\nu'}_{\parallel}(\mathbf{k}'), \mathbf{E}^{\text{sca}}_{\parallel} \rangle - \langle \mathbf{E}^{\nu'}_{\parallel}(\mathbf{k}'), \mathbf{E}^{\text{tra}}_{\parallel} \rangle = 0, \quad (7.16)$$

where the functional dependence of field quantities on the interface coordinates has been dropped for clarity. Substituting the appropriate mode representations then yields

$$\sum_{\nu=s,p} \iint_{\infty} a^{\text{sca},\nu}(\mathbf{k}) \langle \mathbf{E}^{\nu'}_{\parallel}(\mathbf{k}'), \mathbf{E}^{\nu}_{\parallel}(\mathbf{k}) \rangle \, dk_x dk_y - \sum_{\mu} \sum_{m,n} b^{\mu}_{mn} \langle \mathbf{E}^{\nu'}_{\parallel}(\mathbf{k}'), \mathbf{E}^{\mu}_{\parallel mn} \rangle = 0,$$

where the integration is taken over an infinite domain. From Eqs. (B.24) the angular spectra for the s and p polarised plane waves follow as

$$a^{\text{sca},\nu}(\mathbf{k}) = \left[\left(\frac{2\pi}{\omega\mu}\right)^2 \delta_{\nu,s} + \frac{\epsilon}{\mu}\left(\frac{2\pi}{|k_z|}\right)^2 \delta_{\nu,p} \right] \sum_{\mu} \sum_{m,n} b^{\mu}_{mn} \langle \mathbf{E}^{\nu'}_{\parallel}(\mathbf{k}), \mathbf{E}^{\mu}_{\parallel mn} \rangle.$$
$$(7.17)$$

Evaluation of Eq. (7.17) however requires knowledge of the waveguide mode expansion coefficients b^{μ}_{mn}. These coefficients can by found by taking the inner product of Eq. (7.15b) with an arbitrary waveguide mode. Upon further substitution of the appropriate mode representations, in conjunction with Eq. (7.17), the matrix equation

$$\mathbb{A}\mathbf{b} = \mathbf{c} \tag{7.18}$$

results (assuming a suitable lexicographic ordering of the mode indices), where

$$\mathbb{A} = \left[\langle \mathbf{H}_{\|pq}^{\mu'}, \mathbf{H}_{\|mn}^{\mu} \rangle - \frac{4\pi^2}{\omega^2 \mu^2} \iint_\infty \langle \mathbf{E}_\|^s(\mathbf{k}), \mathbf{E}_{\|mn}^{\mu} \rangle \langle \mathbf{H}_{\|pq}^{\mu'}, \mathbf{H}^s(\mathbf{k}) \rangle \, dk_x dk_y \right.$$
$$\left. - \frac{4\pi^2 \epsilon}{\mu} \iint_\infty \langle \mathbf{E}_\|^p(\mathbf{k}), \mathbf{E}_{\|mn}^{\mu} \rangle \langle \mathbf{H}_{|pq}^{\mu'}, \mathbf{H}^p(\mathbf{k}) \rangle \frac{dk_x dk_y}{|k_z|^2} \right], \tag{7.19}$$

$\mathbf{b} = [b_{mn}^{\mu}]$ and $\mathbf{c} = [\langle \mathbf{H}_{\|pq}^{\mu'}, \mathbf{H}_\|^{\text{ill}} + \mathbf{H}_\|^{\text{ref}} \rangle]$. The domain of integration of the inner products of Eq. (7.19) is restricted to the aperture of the data pit since the waveguide modes are zero outside of this domain. Analytic expressions can hence be derived as given in Appendix B.

Inversion of Eq. (7.18) allows the waveguide mode coefficients to be extracted and hence the angular spectrum of the scattered light to be calculated. Characteristic to the modal method presented is that the interaction matrix, $\mathbb{A}$, need only be calculated once for a single pit configuration. Furthermore $\mathbb{A}$ is independent of pit depth. Importantly, this means the depth can be freely changed without requiring further major computational efforts, which is of use in terms of the optimisation of Sect. 7.3. Alternatively the illumination can be varied, as is beneficial for calculation of the polarisation properties as discussed in the next section.

7.2 Optical Disc Readout: Numerical Results

Evaluation of Eqs. (7.17) and (7.18) allows both the field within the data pit and the scattered field to be found. Numerical calculations were conducted in this regard based on the prototype sample described in Sect. 7.1.1, assuming the pit dimensions of [17], i.e. $a = 100$ nm, $b = 25$ nm and $D = \lambda/4$. The results are presented here. Furthermore, propagation of the scattered field through both conventional and confocal readout systems (see Chap. 6) is performed to compare two possible detection architectures. The polarisation characteristics of the readout signal are also presented.

7.2.1 Field Distributions Within the Data Pit

Figure 7.2 shows the individual Cartesian electric field components and optical intensity within the data pit volume, for a pit illuminated by a uniform horizontally (x) polarised beam focused by a lens with NA $= 0.95$ and focal length of 3.5 mm.

7.2 Optical Disc Readout: Numerical Results

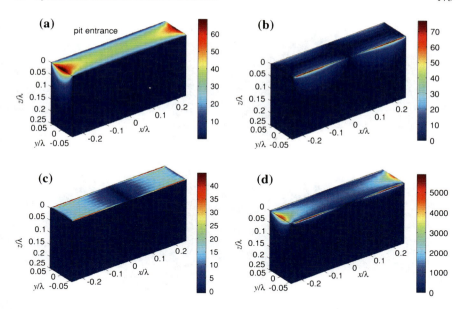

Fig. 7.2 Distributions showing the absolute magnitude of the Cartesian electric field components (E_x^{tra}, E_y^{tra}, E_z^{tra}) (**a**)–(**c**) and corresponding optical intensity (**d**) within a 200 nm × 50 nm × $\lambda/4$ MODS data pit, illuminated with x polarised light of wavelength 405 nm focused by a lens of numerical aperture of 0.95. Maximum waveguide mode indices of $m = n = 30$ were used in calculations. Magnitude scales are in arbitrary units

Waveguide mode expansions were truncated at $m = n = 30$. Similarly, Fig. 7.3 shows the same distributions for a vertically (y) polarised illumination beam.

Coupling strength of the incident light into the pit is dictated by the inner products $\langle \mathbf{E}_{\parallel}^{\nu}(\mathbf{k}), \mathbf{E}_{\parallel mn}^{\mu}\rangle$ and $\langle \mathbf{H}_{\parallel mn}^{\mu}, \mathbf{H}^{p}(\mathbf{k})\rangle$, analytic expressions for which are given in Sect. B.3.5 (see also [2]). Symmetry inherent in these integrals dictate that a normally incident p polarised plane wave (for normal incidence this corresponds to horizontal polarisation) can only couple to TE modes with $m = 0$ and odd n, whilst conversely an s polarised wave (vertically polarised) only excites TE modes with odd m and $n = 0$ [3]. An obliquely incident p (s) polarised plane wave does not suffer this restriction and can couple into arbitrary waveguide modes, however the coupling is weaker for $m \neq 0$ and even n (odd m and $n \neq 0$) orders. Interestingly it is hence expected that y polarised illumination couples more strongly into an x-oriented data pit, in agreement with numerical findings.

Evanescent modes however suffer an exponential decay in amplitude with depth[6] which scales as $|\kappa_z|^{-1}$. Greater penetration depth hence occurs if the incident light couples more strongly into modes with a smaller propagation constant; specifically the $n = 0$ TE modes (since the lowest TM mode is TM$_{11}$) as shown Table 7.1.

---

[6] Due to the backward decaying component arising from reflection off the bottom of the pit the field amplitude, in actuality, decays as $\cosh[|\kappa_z|(z - D)]$ for $0 \leq z \leq D$.

174   7 Multiplexed Optical Data Storage (MODS)

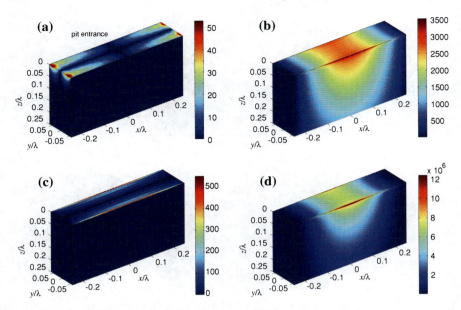

Fig. 7.3 Distributions showing the absolute magnitude of the Cartesian electric field components (E_x^{tra}, E_y^{tra}, E_z^{tra}) (**a**)–(**c**) and corresponding optical intensity (**d**) within a 200 nm × 50 nm × $\lambda/4$ MODS data pit, illuminated with y polarised light of wavelength 405 nm focused by a lens of numerical aperture of 0.95. Maximum waveguide mode indices of $m = n = 30$ were used in calculations. Magnitude scales are in arbitrary units but are comparable to Fig. 7.2

Table 7.1 Propagation constants, in units of $k = 2\pi/\lambda$, for low order field modes in a rectangular waveguide of dimensions 200 × 50 nm

	$m = 0$	$m = 1$	$m = 2$	$m = 3$
$n = 0$	–	0.16 i	1.76 i	2.87 i
$n = 1$	3.92 i	4.05 i	4.42 i	4.96 i
$n = 2$	8.04 i	8.10 i	8.29 i	8.59 i
$n = 3$	12.11 i	12.15 i	12.28 i	12.48 i

Theory, therefore predicts that a y polarised field, coupling predominantly into the TE_{10} mode, produces a larger field amplitude at greater depths in the pit than for a x polarised illumination, which chiefly excites the TE_{01} mode. This feature is clearly apparent in Figs. 7.2 and 7.3.

7.2.2 Properties of the Scattered Field

Having determined the waveguide expansion coefficients the angular spectrum of the scattered light follows, from which either the near or far field can be calculated. The former is more pertinent for readout systems employing a SIL, however the latter is appropriate for considering more traditional readout optics that exist in, say,

7.2 Optical Disc Readout: Numerical Results

a DVD or BD system. Attention is hence restricted to the far field in the following sections. Nevertheless it is worthwhile noting that due to the sub-wavelength size of the MODS pit structures, strong evanescent fields are introduced in the scattered near field suggesting a SIL readout system could present further opportunities.

To describe the potential dependence of the scattered field on the polarisation of the illumination, first recall that each point (θ', ϕ') in the entrance pupil represents an incoming plane wave of strength $\tilde{\mathbf{E}}^{in}$. Each incident plane wave gives rise to a spectrum of scattered plane waves, such that the resulting field, $\tilde{\mathbf{E}}^{out}(\theta, \phi)$ in the exit pupil is given by

$$\tilde{\mathbf{E}}^{out}(\theta, \phi) = \iint_\Theta \tilde{\mathbb{T}}(\theta, \phi, \theta', \phi') \tilde{\mathbf{E}}^{in}(\theta', \phi') \, dS', \qquad (7.20)$$

where Θ denotes the domain of the entrance pupil and $\tilde{\mathbb{T}}(\theta, \phi, \theta', \phi')$ describes the scattering of each incident plane wave component. Assuming a homogeneously polarised illumination, a Jones pupil can be defined by $\iint_\Theta \tilde{\mathbb{T}}(\theta, \phi, \theta', \phi') dS'$. Using the scattering model presented it is possible to calculate the Jones pupil by calculating the field distribution in the back focal plane of the readout lens for x and y polarised illumination separately. As such, the Jones pupil, considering the scattered field, $\mathbf{E}^{sca}$, solely is shown in Fig. 7.4, where the insets also show the Jones pupil when considering the total outward propagating field, $\mathbf{E}^{ref} + \mathbf{E}^{sca}$. Whilst strong preferential transmission is seen for incident y polarised light in the scattered field, due to the relative amplitude of the reflected field this diattenuation is greatly diminished in the total outward field.

Quantification of the diattenuation present however in general depends on the mode of operation of the readout system. A confocal detection scheme, for example, refocuses the field distribution in the pupil on to a point detector, whilst a conventional configuration uses a spatially infinite detector. As such the measured intensity is given by

$$D^{conf} = \left| \mathbb{T} \left[\iint_\Theta \iint_\Theta \tilde{\mathbb{T}}(\theta, \phi, \theta', \phi') \, dS dS' \right] \mathbf{E}^{in} \right|^2 \qquad (7.21)$$

$$D^{conv} = \iint_\Theta \left| \mathbb{T} \left[\iint_\Theta \tilde{\mathbb{T}}(\theta, \phi, \theta', \phi') \, dS' \right] \mathbf{E}^{in} \right|^2 dS \qquad (7.22)$$

where the additional Jones matrix, $\mathbb{T}$, has been introduced to allow for polarimetric measurements (it should be emphasised that $\mathbb{T}$ is not the instrument matrix in this case but is instead the Jones matrix associated with one detection arm in the DOAP used).

Since in the confocal case the measured intensities are dependent on the average field incident on the detector, a confocal polarimeter returns information about the average of Jones pupil. Conventional measurements however record average intensities, or Stokes vectors and hence provide information pertaining to the

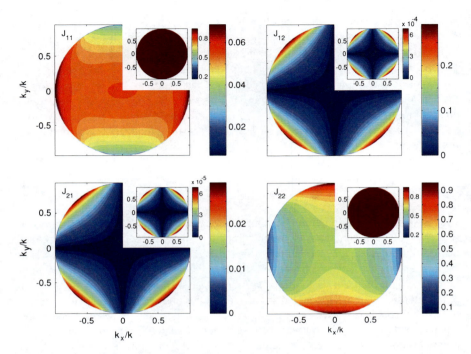

Fig. 7.4 Jones pupil describing the pointwise change in polarisation for the light scattered from the MODS data pit of Figs. 7.2 and 7.3, when illuminated by a focused homogeneously polarised beam. Insets show the Jones pupil for the total outward propagating field. Normalisation is such that the Jones pupil is pointwise passive

average of the Mueller pupil.[7] The latter can hence introduce depolarisation into a measurement even when studying a non-depolarising sample. The Cauchy-Schwarz inequality then implies that significant differences arise in the output signal for the two different readout architectures when the Jones pupil is spatially inhomogeneous. Figure 7.4 however demonstrates this is not the case in the MODS example and little difference is to be expected. Such an expectation is born out upon quantitative analysis, by means of a polar decomposition [16], whereby the parameters shown in Table 7.2 are found. Values for diattenuation, retardance and depolarisation are given for the scattered field alone and the total outward propagating field to highlight the suppression of diattenuation in the latter case. Since little difference is seen between conventional and confocal readout systems, a confocal system shall henceforth be assumed so as to avoid the inherent depolarisation and to allow direct comparison with the results of [17].

In light of the results shown in Figs. 7.2, 7.3, 7.4 and Table 7.2 the origin of the diattenuating properties of a MODS data pit can be deduced. An incident

---

[7] The Mueller pupil is the Mueller calculus equivalent to $\widetilde{\mathbb{T}}(\theta, \phi, \theta', \phi')$ as is more appropriate for a depolarising system.

7.2 Optical Disc Readout: Numerical Results

Table 7.2 Polarimetric parameters as found using a Lu–Chipman decomposition of the simulated readout Mueller matrix

		Diattenuation		Retardance		Depolarisation
		Vector	Norm	Vector	Norm	
Confocal	Total	(1,0,0)	4.3×10^{-4}	(−1,0,0)	0.0012	0
	Scattered	(−1,0,0)	0.9828	(−1,0,0)	0.2977	0
Conventional	Total	(1,0,0)	4.3×10^{-4}	(−1,0,0)	0.0012	∼0
	Scattered	(−1,0,0)	0.9833	(−1,0,0)	0.2973	0.0866

x polarised beam suffers only a weak perturbation from the presence of the pit giving rise to a weak x (and y) field component in the scattered field. On the other hand a y polarised beam couples relatively strongly into the pit and hence the scattered beam has a strong y component (albeit still with a weak x component), hence giving rise to a diattenuation vector of $(-1, 0, 0)$ when the scattered field is considered in isolation. Expressed simply this implies that E_x^{ill} is responsible for any non-zero E_x^{sca} component and similarly for E_y^{ill} and E_y^{sca}.

Upon scattering, however, the transverse components of the scattered field experience different phase shifts (a heuristic description of this process is given in Sect. 7.3). The properties of the total outward propagating field, as given by the coherent superposition of the specularly reflected field and the scattered field are then dictated by these phase differences. Ultimately a larger phase difference between E_y^{sca} and E_y^{ref} gives rise to a greater degree of destructive interference, than in the respective x field components. A strong x component hence results, giving a diattenuation vector of $(1, 0, 0)$.

Recent experiments performed by Macías-Romero [17], to obtain the Mueller pupil associated with readout of a collection of elongated data pits (see Fig. 7.1a), confirm that diattenuation is present in the readout signal, however to a much greater extent than that predicted by the model presented here, with a diattenuation of ∼0.4 being observed in the total field. Reasons for this discrepancy are likely to arise from the assumptions made in the calculations. Suspicion particularly lies with the assumption of a perfect conductor. The silicon substrate used in [17] is far from a perfect conductor, a fact that will greatly affect the field distribution near the pit and the mode coupling strengths. Furthermore, for silicon, with a reflectance of ∼0.4 at 405 nm [22] as compared to a reflectance of unity for a perfect conductor, specular reflection is significantly smaller and hence an increase in diattenuation would be expected. Although the propagation constants of waveguide modes far from cutoff do not differ significantly when a perfectly conducting guide material is assumed relative to a more realistic electrical conductivity [3], the dimensions of the data pit modelled here result in the TE_{10} lying close to the cutoff (see Table 7.1). The disparity in propagation constant in turn can be expected to give rise to a different degree of phase retardation and hence also influence the observed diattenuation.

7.3 Optimal Pit Geometry

Despite the numerical discrepancies between the presented model and experimental results, qualitative results can still be extracted and insight gained into the physical processes determining the form of the readout signal. In particular in this section the geometry of the data pit will be optimised, with a view to maximising the diattenuation present in the total field. By so doing, the region of polarisation space spanned by the output field, for a fixed, linearly polarised illumination is maximised and hence so too is the capacity of the MODS system (c.f. Eq. (5.11)). Optimisation of data pit geometry has previously been considered in the context of detecting an intensity modulation in the readout signal of an ODS system, e.g. [13, 26, 30], however the work presented here in the context of optimisation of polarisation properties is new.

Considering first the effect of altering the depth of a single data pit, it is found that diattenuation in the scattered component of the field increases rapidly from zero to ~ 1 over a short range of pit depths, with complete diattenuation approximately seen for $D > \lambda/5$, as shown by the solid blue curve in Fig. 7.5a. Growth in the total field however occurs over a longer range of depths (see Fig. 7.5b), as a consequence of the slow rate of increase of retardance (dashed green curve) of the scattered field with pit depth.

A striking feature of the results of Fig. 7.5a is the contrast with the behaviour of larger pits. Whilst monotonic increases in both the retardance and diattenuation are seen for data pits of transverse dimensions 200×50 nm, oscillatory variations with D are present for larger pits in which propagating waveguide modes can exist [5, 19]. Figure 7.5c, for example, shows the simulated diattenuation and retardance for a 220×50 nm pit, for which the TE_{10} mode is a propagating mode. Oscillatory behaviour of this nature can heuristically be understood by noting that the phase accumulated by the *propagating* waveguide mode depends on the distance travelled before reflection and hence on the depth of pit. Consequently the interference between the specularly reflected field and that coupled back out of the pit, can be either constructive or destructive in nature. Evanescent modes do not propagate however, implying no phase is accrued and therefore no oscillatory behaviour is exhibited. Instead an analogy can be drawn with total internal reflection and non-propagating waves in transmission lines. In these scenarios the reflection coefficient for waves incident onto an interface has unit amplitude, but is in general complex. Consequently a phase shift is introduced in the reflected beam, albeit a non-oscillatory one.

To corroborate this interpretation a simple transmission line representation was used to model coupling into and reflection from the pit boundary. A transmission line approach is justifiable, since the surface charges and currents in the guide walls introduce a distributed capacitance and inductance, as possessed by electrical transmission lines. The reflection coefficient r between the interface of two transmission lines with differing impedances, Z_1 and Z_2 is a standard result [8] and is thus simply quoted here as

7.3 Optimal Pit Geometry

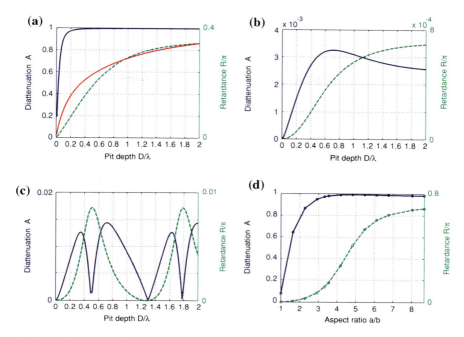

Fig. 7.5 **a** Variation of diattenuation (*blue solid*) and retardance (*green dashed*) seen in the output signal of a confocal MODS readout system, with pit depth, for a pit of transverse dimensions 200 × 50 nm, when considering only the scattered component of the field. *Solid red line* shows the retardance predicted using a simple transmission line analogy (see text). **b** Shows the same as in (**a**) including the specularly reflected field. **c** Gives the variation of diattenuation and retardance as (**a**) assuming a larger pit of transverse dimensions 220 × 50 nm which hence possesses a single propagating waveguide mode. **d** Variation of diattenuation and retardance seen in the output signal of a confocal MODS readout system with transverse aspect ratio for a pit with a fixed depth of $\lambda/4$

$$r = \frac{Z_2 - Z_1}{Z_2 + Z_1}. \tag{7.23}$$

Within the context of the scattering problem discussed in this chapter, the first transmission line, representing region 1, will have impedance equal to that of free space, i.e. $Z_1 = Z_0 = \sqrt{\epsilon_0/\mu_0}$. The impedance of region 2 however depends not only on the type of modes present in the waveguide, but also the strength with which they are present. In particular the characteristic wave impedance for a single waveguide mode is given by [1]

$$Z_{mn}^{TM} = \frac{\kappa_z}{\omega\epsilon}, \qquad Z_{mn}^{TE} = \frac{\omega\mu}{\kappa_z}. \tag{7.24}$$

Equation (7.24) however only accounts for the presence of a single transmitted mode and does not consider the reflected contribution from the back surface of the pit. Again a standard result can be quoted, namely that the impedance of a transmission line of length D, carrying a wave and its reflection due to termination of the line by a load of impedance Z_L, is

$$Z_{mn}^{tot,\mu}(D) = Z_{mn}^{\mu} \frac{Z_L + i Z_{mn}^{\mu} \tan \kappa_z D}{Z_{mn}^{\mu} + i Z_L \tan \kappa_z D}, \qquad (7.25)$$

which reduces to $Z_{mn}^{tot,\mu}(D) = i Z_{mn}^{\mu} \tan \kappa_z D$ upon termination by a perfect conductor ($Z_L = 0$).

In general however multiple modes exist in the data pit and hence the equivalent transmission line must be modelled by a set of impedances in parallel, such that

$$\frac{1}{Z_2} = \sum_{\mu} \sum_{m,n} \frac{b_{mn}^{\mu}}{Z_{mn}^{tot,\mu}}. \qquad (7.26)$$

The retardance in the scattering problem can hence be estimated by calculating the difference in phase of the reflection coefficient, r, for x and y polarised illumination, and is represented by the solid red curve of Fig. 7.5a. Although exact correspondence is not seen due to the oversimplification of the nature of the incident wave, a phenomenological agreement is evident.

Finally, Fig. 7.5d considers the variation of diattenuation and retardance present in the scattered field, as the aspect ratio of the data pit is adjusted. For the calculations b was held constant at a value of 30 nm, whilst a was varied from 30 nm, giving a square pit, to 520 nm. Quite expectably, as the asymmetry is increased the polarisation dependant behaviour increases with an aspect ratio of approximately 1:4 exhibiting near complete diattenuation, well within manufacturable limits.

7.4 Conclusions

Optical data storage systems have shown great practical success, but are however drawing close to their physical limits. A new multiplexed scheme has been presented and modelled in this chapter, in which the polarisation state of light scattered from differently oriented asymmetric pits, is used to represent multiple logical states. The limited polarisation resolution of any practical polarimetric readout system however limits the number of distinguishable states. Maximising the capacity of a polarisation multiplexed system therefore entails maximising the domain of possible scattered states for a fixed illumination. Optimisation of this nature however requires a detailed knowledge of the polarisation dependant properties of the system as a whole, hence necessitating a suitable scattering model to be constructed.

The large number of calculations inherent to optimisation motivated a modal approach to the scattering problem, in which the fields above the disc surface and within the data pit volume were decomposed into a physically complete set of indexed solutions to Maxwell's equations. Scattering calculations were thus reduced

7.4 Conclusions

to calculation of a single interaction matrix. Inversion of the resulting matrix equation allowed the field distributions within the pit, in the near and far field to be calculated.

Since the interaction matrix is independent of the incident field, illumination conditions could be altered with limited additional computational effort, allowing the Jones pupil to be calculated. This was performed assuming a spatially homogeneous polarised beam. Given the calculated Jones pupil the polarisation properties of the output signal for both a confocal and a conventional setup were considered. Little difference was seen between the two configurations due to the homogeneity of the field in the pupil plane, however both exhibited strong diattenuation in the scattered field. Diattenuation was argued to arise due to differing phase delays introduced into the scattered x and y field components. Heuristically this behaviour was described by drawing an analogy with coupling to evanescent waves in transmission line theory. In this model the effective impedance of each evanescent mode within the pit domain varies with the propagation constant, such that the illumination dependent mode coupling produces a retardance in the scattered field and hence diattenuation in the total field.

Evanescent coupling in sub-wavelength pit structures was also seen to give rise to significantly different behaviour to existing ODS systems which exhibit oscillatory variations in polarisation parameters with pit depth. Instead the MODS pits considered displayed an increasing polarisation dependence in both the scattered and total field as the pit depth was increased. Asymptotic limits were however seen, for example diattenuation in the total field draws close to its maximum value of unity for a pit approximately one wavelength deep. Again this is in contrast to intensity modulated systems in which the optimal pit depth is an odd integer of quarter wavelengths [10]. Diattenuation was also shown to increase as the transverse aspect ratio of the pit opening was increased.

A number of simplifying assumptions were made in the simulations presented in this chapter, notably that of perfect conductivity. Such assumptions present a limit to the validity of the quantitative findings, with disparities between simulations and recent experimental results being discussed. In particular material differences are expected to play a key role in explaining these disparities, but are however currently not fully understood. Relaxation of the assumption of perfect conductivity remains as future work. Qualitative results are however expected to hold since the underlying physical processes will remain the same in more realistic models.

References

1. C.A. Balanis, *Advanced Engineering Electromagnetics* (Wiley, New York, 1989)
2. J.M. Brok, An analytic approach to electromagnetic scattering problems. Ph.D. thesis, 2007
3. J.M. Brok, H.P. Urbach, A mode expansion technique for rigorously calculating the scattering from 3D subwavelength structures in optical recording. J. Mod. Opt. **51**, 2059–2077 (2000)
4. J.M. Brok, H.P. Urbach, Simulation of polarization effects in diffraction problems of optical recording. J. Mod. Opt. **49**, 1811–1829 (2002)

5. X. Cheng, H. Jia, D. Xu, Vector diffraction analysis of optical disk readout. Appl. Opt. **39**, 6436–6440 (2000)
6. J. Chon, P. Zijlstra, M. Gu, Five-dimensional optical recording mediated by surface plasmons in gold nanorods. Nature **459**, 410–413 (2009)
7. K. Şendur, W. Challener, C. Peng, Ridge waveguide as a near field aperture for high density data storage. J. Appl. Phys. **96**, 2743–2752 (2004)
8. W.J. Duffin, *Electricity and Magnetism* (W. J. Duffin Publishing, Yorkshire, 2001)
9. Y. Honguh, Readout signal analysis of optical disk based on approximated vector diffraction theory. Jpn. J. Appl. Phys. **42**, 735–739 (2003)
10. A. Huijser, J. Pasman, G. van Rosmalen, K. Schouhamer, I.G. Bouwhuis, J. Braat, *Principles of Optical Disc Systems* (Adam Hilger, Bristol, 1985)
11. K. Kobayashi, Vector diffraction modeling: polarisation dependence of optical readout/servo signals. Jpn. J. Appl. Phys. **32**, 3175–3184 (1993)
12. A.S. Lapchuk, A.A. Kryuchin, V.A. Klimenko, Three dimensional vector diffraction analysis for optical disc. Proc. SPIE **3055**, 37–42 (1997)
13. Y. Li, C. Mecca, E. Wolf, Optimum depth of the information pit on the data surface of a compact disk. J. Mod. Opt. **50**, 199 (2003)
14. J. Liu, B. Xu, T.C. Chong, Three-dimensional finite-difference time-domain analysis of optical disk storage system. Jpn. J. Appl. Phy. **39**, 687–692 (2000)
15. W.-C. Liu, Vector diffraction from subwavelength optical disk structures: two-dimensional near-field profiles. Opt. Express **2**, 191–197 (1998)
16. S.Y. Lu, R.A. Chipman, Interpretation of Mueller matrices based on polar decomposition. J. Opt. Soc. Am. A **13**, 1106–1113 (1996)
17. C. Macías Romero, High numerical aperture Mueller matrix polarimetry and applications to multiplexed optical data storage. Ph.D. thesis, 2010
18. S.M. Mansfield, G.S. Kino, Solid immersion microscope. Appl. Phys. Lett. **57**, 2615–2616 (1990)
19. D.S. Marx, D. Psaltis, Polarization quadrature measurement of subwavelength diffracting structures. Appl. Opt. **36**, 6434–6440 (1997)
20. M.G. Moharam, T.K. Gaylord, Rigorous coupled-wave analysis of planar-grating diffraction. J. Opt. Soc. Am. **71**, 811–818 (1981)
21. M.G. Moharam, T.K. Gaylord, Rigorous coupled-wave analysis of grating diffraction—e-mode polarization and losses. J. Opt. Soc. Am. **73**, 451–455 (1983)
22. E.D. Palik (ed.), *Handbook of Optical Constants of Solids* (Elsevier Academic Press, New York, 1998)
23. A. Roberts, Electromagnetic theory of diffraction by a circular aperture in a thick, perfectly conducting screen. J. Opt. Soc. Am. A **4**, 1970–1983 (1987)
24. C. Rockstuhl, H.P. Herzig, Calculation of the torque on dielectric elliptical cylinders. J. Opt. Soc. Am. A **22**, 109–116 (2005)
25. A. Sommerfeld, *Partial Differential Equations in Physics* (Academic Press, New York, 1949)
26. S. Stallinga, G. Hooft, J. Braat, Comment on 'Optimum depth of the information pit on the data surface of a compact disk'. J. Mod. Opt. **51**, 775–778 (2004)
27. Technical University of Delft. Super laser array memory. Final Report IST-2000-26479, International Telecommunication Union, 1998.
28. P. Török, P. Varga, Electromagnetic diffraction of light focused through a stratified medium. Appl. Opt. **36**, 2305–2312 (1997)
29. X. Wei, A.J. Wachters, H.P. Urbach, Finite-element model for three-dimensional optical scattering problems. J. Opt. Soc. Am. A **24**, 866–881 (2007)
30. Z. Zhou, Y. Ruan, Optimization of information pit shape and read-out system in read-only and write-once optical storage systems. Appl. Opt. **27**, 728–731 (1988)
31. F. Zijp, M.B. van der Mark, J.I. Lee, C.A. Verschuren, B.H.W. Hendriks, M.L.M. Balisteri, H.P. Urbach, M.A.H. van der Aa, A.V. Padiy, Near field read-out of a 50 GB first-surface disc with NA = 1.9 and a proposal for a cover-layer incident, dual-layer near field system. Proc. SPIE **5380**, 209–223 (2004)

Chapter 8
Single Molecule Studies

> *A physicist is just an atom's way of looking at itself.*
> Niels Bohr

8.1 Introduction

Single molecule detection (SMD) has become an important technique in recent years for studying dynamic processes such as chemical reactions and molecular motions at a fundamental level [17, 30]. Historically these processes are usually studied using methods based on ensemble averaging of a sample of molecules, however frequently the mean properties so found are insufficient. Studies on single molecules are thus advantageous as information, such as statistical distributions of particular quantities, is not lost by averaging. It should however be noted that even single molecule studies yield results that are temporally averaged over the course of the finite measurement time.

Single molecule imaging techniques, such as fluorescence microscopy, can also be used to track bio-molecular motions. This has applications in the pharmaceutical industry where a good understanding of processes such as protein folding [28] and molecule motions [16] is vital to new drug development. Of particular interest is the determination of the orientation of the emission dipole of single molecules since it can be used as a means to label biological structures and track their conformational changes and motions [2, 14, 29]. Furthermore, photophysical parameters of fluorophores, such as fluorescence lifetime, can depend on the molecule's orientation, a fact which can be used to study the molecule itself or its environment [4, 10].

Optical techniques in single molecule imaging however almost always require the use of photon counting since individual fluorescent molecules are very weak light sources. Under these conditions the accuracy of measurements are limited by random variations in the measured signal, and statistical processing must thus often be

M. R. Foreman, *Informational Limits in Optical Polarimetry and Vectorial Imaging*,
Springer Theses, DOI: 10.1007/978-3-642-28528-8_8,
© Springer-Verlag Berlin Heidelberg 2012

used to extract the desired information. This however requires a good understanding of the random processes present. Poisson statistics, arising from the quantisation of light, has been discussed heavily throughout this thesis, however random fluctuations in the light source can prove important, as will be illustrated in Sect. 8.2.1. In what follows consideration will not be given to antibunching of photons that occurs when considering single photon sources. Instead results can be considered as average results over a number of such single source studies, or over multiple sources.

In this chapter, one particular source of signal fluctuations pertinent to orientational measurements is considered, namely those arising from random rotations which fluorophores may undergo; a phenomenon that shall be referred to as "wobbling". Wobble of fluorophores, which within the framework of classical electrodynamics can be considered as electric dipole emitters, can be either a continuous angular variation or discrete orientational jumps [11], both of which will be considered in the following and seen to possess different statistics (Sect. 8.2.2). Successive jumps in the latter case may furthermore depend on previous dipole orientations and the consequences when this is true and when it is not are discussed. Limiting forms for both slow and fast wobble are also derived.

The latter half of this chapter (Sect. 8.3) considers a further aspect of single molecule orientational measurements and presents a novel technique developed by the author in collaboration with Macías Romero [7], which addresses the experimental difficulties in measuring the longitudinal component of an electric dipole moment. Although methods based on structured illumination, image fitting and total internal reflection [5, 19, 22] exist, in which the signal depends upon the full three dimensional orientation of a dipole, they are often restricted to specific circumstances, can be subject to poor SNRs and are not suitable for real time measurements. It is shown that by breaking the symmetry of the back focal plane field distributions arising when imaging an electric dipole it is possible to introduce a dependence on the longitudinal component of its dipole moment to the on-axis image field. More specifically, a scheme wherein one half of the collected beam is subject to a π phase delay is presented. Potential sources of error for experimental implementations are also discussed in Sect. 8.3.2. Following the underlying theme of this thesis it is seen that a technique capable of measuring the full 3D moment of a dipole (and hence also all three components of an illuminating electric field vector) allows a further information channel to be analysed and exploited in electromagnetic systems.

8.2 Photon Statistics in Single Molecule Orientational Imaging

8.2.1 Signal-to-Noise Considerations

Statistical fluctuations in the number of detected photons can originate either from noise present in a system or from random variations in the signal itself. The relative importance of these sources can be seen by considering the SNR obtained considering

8.2 Photon Statistics in Single Molecule Orientational Imaging

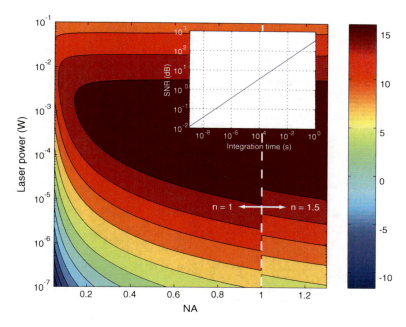

Fig. 8.1 Contour plot of SNR (dB) versus laser power and focused spot size (as parameterised by the NA of a focusing lens) assuming a wavelength of 395 nm and the following parameter values based on use of green fluorescent protein (GFP): $F = 7\%$, $q = 0.79$ [20], $t_0 = 0.01$ s, $C_b = 2 \times 10^8$ photons/Ws, $n_b = 50$. For numerical apertures greater than unity an oil immersion lens of refractive index 1.5 was assumed. Saturation effects are included such that $\sigma = \sigma_0/(1+I/I_s)$, where σ_0 was taken to be 4×10^{-16} cm^2, $I = P/A$ and the saturation intensity I_s was assumed to be 10^3 W/cm^2 [21]. The *inset* shows the variation of the SNR versus integration time for a 5 mW laser focused through a 0.95 NA lens

only the quantisation of light. In the context of single molecule imaging, Basché [1] states that the practically obtainable SNR can be approximated by

$$\text{SNR} = \frac{F\,q\sigma\,P\,t_0/A\,E_p}{\sqrt{(F\,q\sigma\,P\,t_0/A\,E_p) + C_b P\,t_0 + n_b t_0}}, \qquad (8.1)$$

where F is an instrument dependent collection factor typically ranging from 1–8%, q is the fluorescence quantum yield, σ is the peak absorption crosssection of the molecule being studied, P is the laser power, t_0 is the integration time, A is the beam area, E_p is the energy of a photon in the beam, C_b is the background count per Watt of excitation power (typically around 2×10^8 photons/Ws in confocal experiments) and n_b is the dark count of the detector. Figure 8.1 shows the behaviour of the SNR over a range of experimental conditions from which it can be seen that a value no better than around 15 dB is to be expected. Since this value is so low, variations in the signal source can play an equally important role, if not a dominant one, when determining the statistical behaviour of the detected signal. One such noise source in orientational measurements is thus considered and its relative importance in the detection process ultimately evaluated in the subsequent sections.

8.2.2 Probability Density Function of the Number of Detected Photons

Data acquisition in single molecule experiments is invariably done by means of photon counting in which the predominant source of noise is quantisation noise. Denoting the number of photons arriving at the detector during a measurement of duration t_0 by $N(t_0)$, the output reading is of the form $D_{out} = GN$, where G is some gain factor. The arrival of photons at the detector is a Poisson random process with PDF[1] (c.f. Eq. (2.46))

$$p_N(n) = \frac{[R(t_0)]^n}{n!} \exp[-R(t_0)]. \qquad (8.2)$$

R is the average rate of arrival of photons (intensity in units of $h\nu$) or equivalently the time average of the instantaneous rate of arrival of photons at the detector, $\mathcal{R}(t)$ (c.f. Eq. (2.47)), i.e.

$$R(t_0) = \int_0^{t_0} \mathcal{R}(t) dt, \qquad (8.3)$$

where the functional dependence on t_0 will henceforth be suppressed for clarity.

As an example many experimental setups use polarisation sensitive methods [9] whereby the intensity of the detected signal is proportional to the square of the dot product of the illuminating field and the electric dipole moment giving

$$\mathcal{R}(t) = A \cos^2(\gamma(t) - \beta), \qquad (8.4)$$

where $\gamma(t)$ is the transverse orientation of the dipole at time t, β is the transverse angle of the plane of polarisation of incident light and A is a constant.

For a stationary dipole, Eq. (8.2) fully describes the photon statistics at the detector, however a change in dipole orientation will cause a change in $\mathcal{R}$. If this change is random the arrival of photons at the detector, and hence their subsequent detection, is termed a doubly stochastic process. Possible sources of such randomness include fluctuations in the illuminating light source and/or movement of the molecule. Since for tracking applications the molecule's environment is unlikely to be static, it is this latter factor that shall be studied here. Furthermore, only orientational changes will be considered since probe molecules are often rigidly fixed to targets. Under these circumstances R is a random variable and the probabilities as given by Eq. (8.2)

---

[1] The convention whereby an upper case letter denotes a random process and/or variable, whilst the lower case equivalent denotes a particular outcome is again used throughout this chapter.

8.2 Photon Statistics in Single Molecule Orientational Imaging

differ for each possible value. As such Eq. (8.2) is recast as a conditional PDF, as in Sect. 2.1.3.1, where r will be used to denote a particular outcome of R.

Assuming knowledge of the random nature of the time averaged photon rate, as characterised by its PDF $f_R(r)$ (see Sect. 8.2.3), Eq. (2.18) can be used to find the joint PDF of N and R, i.e. the probability that $N = n$ and $R = r$. Integrating over the joint PDF gives the marginal PDF of the number of detected photons

$$p_N(n) = \int_0^\infty f_R(r) \frac{(\eta r)^n}{n!} e^{-\eta r}\, dr. \tag{8.5}$$

where the non-ideal nature of the detector has also been included by introduction of the quantum efficiency η. Equation (8.5), known as Mandel's formula, is equivalent to averaging the conditional probability with respect to the average intensity and requires knowledge of $f_R(r)$ which is discussed in the following section.

8.2.3 Probability Density Function of Time Averaged Intensity

8.2.3.1 Discrete Reorientational Jumps

In this section attention is given to determining the PDF of the time averaged intensity $f_R(r)$. The case when changes in the orientation of a dipole occur discretely is considered first. This could for example be associated with the desorption and readsorption of fluorophores from and onto a glass surface [9]. In what follows, an electric dipole will be said to be in an orientational state, by which it is meant that the dipole makes an angle γ to the x-axis in the x-y plane and an angle χ to the z-axis, as illustrated in Fig. 6.13b. The dipole then remains fixed at this angle for a time τ before moving to a new state. Here discussion will be restricted to a two dimensional system (i.e. $\chi = \pi/2$) for simplicity, but also because the output signal in many experimental techniques is only sensitive to the transverse angle γ (c.f. Eq. (8.4)). Conceptually the full three dimensional situation is identical and requires only minor mathematical modifications as is discussed in Sect. 8.2.4.

Assuming that M different orientational states are occupied during a single measurement the time averaged photon rate is given by:

$$r = A\left(\cos^2(\gamma_1 - \beta)\tau_1 + \cos^2(\gamma_2 - \beta)\tau_2 + \cdots + \cos^2(\gamma_M - \beta)\tau_M\right), \tag{8.6}$$

where γ_j and τ_j are the parameters corresponding to the jth occupied angular state. Without loss of generality, the dipole is assumed to be initially orientated parallel to the x-axis. It should be further noted that changes in the dipole angle are assumed to occur instantaneously.

It is assumed that the law of rare events is applicable such that M is a Poisson random variable. Expressed alternatively, albeit equivalently, the length of time a dipole remains in each state is distributed according to an exponential law [13], i.e.

$$f_T(\tau) = v \exp(-v\tau) \tag{8.7}$$

where $f_T(\tau)$ denotes the PDF of τ, and v is the average rate at which dipole jump events occur.

To approach the problem, first assume a fixed M and let $X_j = A\cos^2(\Gamma_j - \beta)$ and $Z_j = X_j T_j$ such that $R = \sum_{j=1}^{M} Z_j$ and

$$f_{Z_j}(z_j) = \int_0^\infty \int_0^A \delta(z_j - x_j \tau_j) f_{X_j, T_j}(x_j, \tau_j) dx_j d\tau_j. \tag{8.8}$$

Since a measured intensity is always positive the Laplace transform can be defined for each intensity contribution, Z_j, in Eq. (8.6) (see Sect. 2.1.1.2), whereby

$$Z_j^*(s) = \int_0^\infty f_{Z_j}(z_j) \exp(-sz_j) dz_j, \tag{8.9}$$

which from Eq. (8.8) can be written in the form

$$Z_j^*(s) = \int_{-\pi}^{\pi} \int_0^\infty f_{\Gamma_j, T}(\gamma_j, \tau) \exp\left(-sA\cos^2(\gamma_j - \beta)\tau\right) d\tau\, d\gamma_j. \tag{8.10}$$

It should be noted that in Eq. (8.10) the subscript j on T and τ has been dropped since each τ_j term is assumed to be identically and independently distributed.[2] Following [15] and applying Eq. (8.10) the Laplace transform of $f_R(r|m)$ is then given by

$$R_M^*(s) = Z_1^*(s) Z_2^*(s) \cdots Z_M^*(s), \tag{8.11}$$

from which the PDF of the average photon rate R then follows by performing the expectation with respect to M such that:

$$f_R(r) = \sum_{m=0}^{\infty} p_M(m) f_R(r|m), \tag{8.12}$$

where $f_R(r|m) = \mathcal{L}^{-1}\left(R_{M=m}^*(s)\right)$ and the weighted summation over the possible values of $M = m$ is required since the number of reorientations during a measurement is random.

With this knowledge in hand it remains to find an explicit expression for $Z_j^*(s)$. Since dipole angle and state occupancy time are assumed independent such that $f_{\Gamma_j, T}(\gamma_j, \tau)$ is given by the product of the marginal probability distributions $f_{\Gamma_j}(\gamma_j)$

---

[2] The independence of each Z_j can be shown to follow from the assumed independence of τ_j regardless of the independence of γ_j [15].

8.2 Photon Statistics in Single Molecule Orientational Imaging

and $f_T(\tau)$, Eq. (8.10) can be evaluated. Using Eq. (8.7) it follows that

$$Z_j^*(s) = \int_{-\pi}^{\pi} \int_0^{\infty} f_{\Gamma_j}(\gamma_j) \nu \exp\left(-\nu\tau - sA\cos^2(\gamma_j - \beta)\tau\right) d\tau\, d\gamma_j,$$

$$= \int_{-\pi}^{\pi} \frac{\nu f_{\Gamma_j}(\gamma_j)}{\nu + sA\cos^2(\gamma_j - \beta)} d\gamma_j. \tag{8.13}$$

The physical process governing the random wobble of the electric dipole will dictate the form of the probability distribution for Γ_j. For example, rebinding of a fluorophore to a probe site may be modeled using a uniform PDF

$$f_{\Gamma_j}^{\text{uni}}(\gamma_j) = \begin{cases} 1/2\Delta & \text{for } -\Delta \leq \gamma_j < \Delta \\ 0 & \text{otherwise} \end{cases}. \tag{8.14}$$

Standard integration tables [8] then give the analytic result

$$Z_{\text{uni}}^*(s) = \frac{1}{2\Delta\sqrt{sA\nu + \nu^2}} \left[\arctan\left(\sqrt{\frac{\nu}{sA+\nu}} \tan(\Delta - \beta)\right) \right.$$
$$\left. + \arctan\left(\sqrt{\frac{\nu}{sA+\nu}} \tan(\Delta + \beta)\right) \right]. \tag{8.15}$$

Finding a full analytical result for $f_R(r)$ is complicated however in the limits of small and large ν simpler results naturally emerge. These limits correspond to only a few, and to many events per measurement respectively. As the rate at which jump events occur decreases the contribution from high m terms in Eq. (8.12) becomes negligible. In the limit of $\nu \ll 1$ only the first term produces a significant contribution and the dipole can be considered as fixed during a single measurement and hence

$$f_R(r) = f_{\mathcal{R}}(\mathcal{R}), \tag{8.16}$$

i.e. the PDF of the average intensity is the same as the PDF for the instantaneous intensity. Fortunately this agrees with intuitive expectations.

When dipole wobble is on a time scale much shorter than the duration of a measurement, i.e. large ν, many terms in the summation of Eq. (8.12) must be considered. Since each value of τ_j is independent, each Z_j term is also independent. There are then two cases to consider; that when each subsequent value of γ is independent and that when they are not. In the former case the Central Limit Theorem can be invoked. As such the PDF of the average intensity in the limit of large ν is given by

$$f_R(r) = \frac{1}{\sqrt{2\pi\sigma_R^2}} \exp\left(-\frac{r^2}{2\sigma_R^2}\right). \tag{8.17}$$

Assuming dependence of consecutive terms means the PDF of the dipole angle γ_j is centered on its previous outcome, γ_{j-1}. For a particular realisation of γ, that is to say one possible outcome of the sequence of dipole orientations,

$$f_{\Gamma_j}(\gamma_j) = f_\Gamma(\gamma - \gamma_{j-1}). \tag{8.18}$$

When averaged over all possible realisations the result is similar to Eq. (8.14) except now the width of the distribution increases with each subsequent jump. Consequently the condition of identical distributions required for validity of the Central Limit Theorem is not satisfied. If, however, the Lyapunov condition [6] is satisfied then the Central Limit Theorem still applies. Numerical simulations show that this is the case.

8.2.3.2 Continuous Angular Variation

Changes in dipole orientation may occur continuously and it is here that consideration is given as to how this affects the PDF of the time averaged intensity. It can be shown [3] that the probability distribution function of the orientation of the dipole at a time t satisfies the differential equation:

$$\frac{\partial f_\Gamma}{\partial t} = \alpha \frac{\partial^2 f_\Gamma}{\partial \gamma^2}, \tag{8.19}$$

subject to the initial condition $f_\Gamma(t = 0) = \delta(\gamma - \gamma_0)$, where δ represents the Dirac delta function. This diffusion equation holds when subsequent orientations are dependent on the previous orientation. A solution to Eq. (8.19) is

$$f_\Gamma(\gamma, t) = \frac{1}{\sqrt{4\pi\alpha t}} \exp\left(-\frac{(\gamma - \gamma_0)^2}{4\alpha t}\right), \tag{8.20}$$

where an implicit assumption has been made that time intervals and diffusion rates (as set by the diffusion coefficient α) are small enough such that the PDF has not been equalised over all angles.

To find the PDF of the average intensity, a transformation of variables is first used to find the PDF of the instantaneous intensity $f_\mathcal{R}(\mathcal{R})$ via Eq. (2.21) which, upon integration over the length of a measurement, gives the desired result (see Sect. 2.1.2). Thus

$$f_R(r) = \frac{1}{t_0} \sum_k \int_0^{t_0} \frac{f_\Gamma(\gamma_k, t)}{\sqrt{r(A-r)}} dt, \tag{8.21}$$

where γ_k are the solutions to the equation $r = A\cos^2(\gamma - \beta)$ and the $1/t_0$ factor is to ensure correct normalisation of the PDF. The integral can be evaluated using the substitution $x^2 = t^{-1}$ and integration by parts which yields:

8.2 Photon Statistics in Single Molecule Orientational Imaging

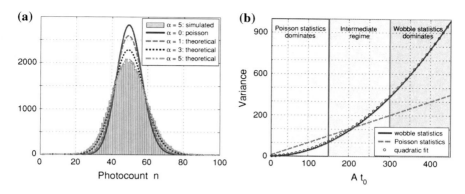

Fig. 8.2 a Histogram of the time averaged intensity for a dipole undergoing continuous angular diffusion with $\alpha = 5$, $\beta = \pi/4$, $t_0 = 10^{-3}$ s and $A = 10^5$ photons/s shown with theoretical fits for different diffusion coefficients; b variance of the number of detected photons as a function of the peak signal strength A

$$f_R(r) = \frac{1}{\sqrt{\pi \alpha t_0 \, r(A-r)}} \sum_k \left[\exp\left(-\frac{\gamma_k^2}{4\alpha t_0}\right) - \frac{|\gamma_k|}{2\alpha t_0} \operatorname{erfc}\left(\sqrt{\frac{\gamma_k^2}{4\alpha t_0}}\right) \right], \tag{8.22}$$

where erfc(...) denotes the complimentary error function.

For the independent case $f_\Gamma(\gamma, t)$ can not depend on time (assuming the physical cause of the wobble does not vary in time) and as such Eq. (8.21) reduces to $f_R(r) = f_\mathcal{R}(\mathcal{R})$.

Figure 8.2a shows a histogram of the results of Monte-Carlo simulations with 10^4 realisations for continuous variation and a diffusion coefficient of $\alpha = 5$. Various theoretical fits, as based on Eq. (8.22), are also drawn from which it can be seen that for $\alpha = 0$ (no dipole wobble) the PDF is identical to that of a Poisson distribution as would be expected. Good agreement can also be seen between the simulated and theoretical results.

Furthermore, using these PDFs it is possible to calculate the total cumulative probability of N taking any value below n. Confidence levels including or neglecting dipole wobble can then be calculated. Assuming the values $\beta = \frac{\pi}{4}$, $\gamma_0 = 0$, $A = 10^5$ photons/s, $t_0 = 10^{-3}$ s and $\alpha = 5$ it was calculated that when neglecting dipole wobble an experimental measurement can determine the orientation of a dipole within a range of 1.78° with 90% confidence. Inclusion of dipole wobble causes this to increase to 2.43°.

Discrepancies such as that seen in the previous calculation further highlights the need to include dipole wobble in statistical processing and error analysis where appropriate. In this regard the reader's attention is drawn to Fig. 8.2b which shows a plot of the expected variance of the photon count n as a function of the number of photons in the system (as parameterised by A), when signal variations from photon counting and dipole wobble are considered separately.[3] Quadratic behaviour can be seen for the case of dipole wobble only, whilst for photon counting the linear behaviour expected from a pure Poisson random variable is evident. The relative importance of the two factors can be seen. At very low light intensities, where it is likely to be impractical to conduct experiments, photon counting dominates. For the intermediate regime both influences are comparable until eventually at higher intensities the molecular wobble dominates.

8.2.4 Three Dimensional Dipole Wobble

Earlier discussion was restricted to the case of two dimensional dipole wobble, however here the mathematical modifications required to accommodate rotation in all three dimensions are explicitly given. Obviously such three dimensional variation is only of significance if the measured signal is sensitive to the full three dimensional orientation as described by the two angles γ and χ as shown in Fig. 6.13, i.e.

$$\mathcal{R} = \mathcal{R}(\gamma, \chi). \tag{8.23}$$

Random variation in the orientation of the dipole is described by the joint PDF of Γ and χ. Since in most physical situations Γ and χ will be independent the joint PDF can be written in the form

$$f_{\Gamma,\chi}(\gamma, \chi) = f_\Gamma(\gamma) f_\chi(\chi). \tag{8.24}$$

When considering the discrete case this means Eq. (8.10) becomes a triple integral

$$Z_j^*(s) = \int_{-\pi}^{\pi} \int_0^{\pi} \int_0^{\infty} f_\Gamma(\gamma) f_\chi(\chi) f_T(\tau) \exp\left(-s\mathcal{R}(\gamma, \chi)\tau\right) d\tau \, d\chi \, d\gamma, \tag{8.25}$$

however all of the subsequent working remains unchanged. In the continuous case the two dimensional diffusion equation must be solved

$$\frac{\partial f}{\partial t} = \alpha \left(\frac{\partial^2 f}{\partial \gamma^2} + \frac{\partial^2 f}{\partial \chi^2} \right), \tag{8.26}$$

to give the joint PDF

$$f_{\Gamma,\chi}(\gamma, \chi) = \frac{1}{4\pi\alpha t} \exp\left(-\frac{(\gamma - \gamma_0)^2}{4\alpha t}\right) \exp\left(-\frac{(\chi - \chi_0)^2}{4\alpha t}\right), \tag{8.27}$$

---

[3] Although arguably a Fisher information analysis, as has been extensively utilised in this text, is possible using the derived PDFs, this is omitted here since consideration of the variance is adequate in highlighting the key points.

8.2 Photon Statistics in Single Molecule Orientational Imaging

which when integrated according to the three dimensional analogue of Eq. (8.21) yields

$$f_R(r) = \frac{1}{8\pi\alpha t_0 \sqrt{r(A-r)}} \sum_k \left[\widetilde{\Gamma}\left(0, \frac{(\gamma_k - \gamma_0)^2 + (\chi_k - \chi_0)^2}{4 t_0 \alpha}\right)\right], \quad (8.28)$$

where $\widetilde{\Gamma}(a, z) = \int_z^\infty x^{a-1} e^{-x} dx$ is the incomplete Gamma function and γ_k and χ_k are the solutions to the equation $r = \mathcal{R}(\gamma, \chi)$.

8.2.5 Discussion

It has been shown that the variation of the orientation of a dipole over the course of a finite duration measurement can alter the statistical properties of the number of photoelectrons induced in a photon counting detector. Although analytic evaluation of Eqs. (8.5) and (8.12) will in general not be possible, some general observations can be made regarding the probability functions involved.

Considering first the PDF of the time averaged photon rate, different forms and behaviour for differing dependence conditions and time scales can be expected. More specifically, although the distribution will always be peaked around the initial angle, when subsequent orientations of the dipole are dependent on earlier positions the distribution is narrower for slower changes, whilst the converse is true when independence holds. This can be understood since the dependent situation is essentially a diffusion problem and so the larger the ratio of spreading rate (as given by the diffusion coefficient) to integration time the larger the range of angles the dipole can cover during a measurement. On the other hand the distribution focuses when successive orientations are independent since the central probability peak for each Z term is reinforced with each additional term in the average.

For wobbling on faster time scales the PDF of the average intensity has been shown to tend to that of a bell-shaped Gaussian distribution (for discrete variations). Slower time scales will exhibit a sharper more centralised distribution, since for small ν, i.e. slow variation, only a few terms significantly contribute to the average performed by the detector. In this case the peaked nature of both the exponentially distributed state occupancy times and the Poisson PDF for the number of events per measurement dominate. For larger ν the Poisson PDF becomes smoother and the position of the peak moves to larger m. Low m terms of Eq. (8.12) are then negligible and the peaked nature of the exponential PDF is less dominant. Eventually the Poisson PDF tends to a Gaussian itself whereby it acts as an envelope for the PDF of the average dipole angle.

In terms of the photoelectron statistics, it can be said that for smaller angular ranges of dipole oscillation one would expect less deviation from conventional Poissonian behaviour. Furthermore, if the variations are on a timescale much longer than the integration time then the additional random behaviour will be unobservable. On the other hand if fluctuations are much faster than the detector response the effects are likely to again go unnoticed. That said, dipole wobble has been seen at many different time scales ranging from the subnanosecond level [27], through the millisecond regime [14] and higher [11, 31]. In conjunction with the varying time resolution of different experimental setups [17] and the large angular ranges over which fluorophores can oscillate, e.g. 26° has been observed [18], it is likely that non-Poisson behaviour will be frequently encountered.

8.3 Longitudinal Dipole Orientation and Field Mapping

Many current techniques for determining the orientation of a dipole exist e.g. [9, 14] however these are commonly limited to finding the transverse angle. Uncertainty over the longitudinal dipole component can lead to large errors (e.g. [10]) and hence it is desirable to find the full 3D orientation of the molecule. In this section a novel technique capable of determining the 3D orientation of a single fluorescent molecule in real time is thus presented. As is pertinent to any practical technique, the experimental tolerances of the setup are investigated including misalignments and the finite width of the molecule's emission spectrum.

8.3.1 Description of System

Before tackling the issue of how the longitudinal orientation of a fluorescent molecule (again modelled as an electric dipole emitter with moment $\mathbf{p} = (p_x, p_y, p_z)$) can be measured it is insightful to revisit the topic of imaging a dipole as discussed in Sect. 6.4.2.1. Therefore, once more consider collecting and collimating the far field radiation pattern of an electric dipole using a high numerical aperture, aplanatic lens, assumed to be ideal and immersed in a medium of the same refractive index as that containing the dipole. Although this constitutes a special case, a fuller treatment in which the dipole field also propagates through a number of dielectric interfaces (e.g. a cover glass-immersion fluid interface) can be modelled using the theory detailed in [23], however the symmetry of the problem is unaltered and hence the following discussion is also applicable. It is also worthwhile to note that, although electric dipoles will be exclusively considered in what follows, the discussed technique works equally well when imaging magnetic dipoles. Vectorial ray tracing can be used to find the electric field $\widetilde{\mathbf{E}}$ in the back focal plane of the collector lens as was detailed in Sect. 6.4.2.1. Repeating the result here for ease of reference the collimated field collected from an electric dipole is given by

8.3 Longitudinal Dipole Orientation and Field Mapping

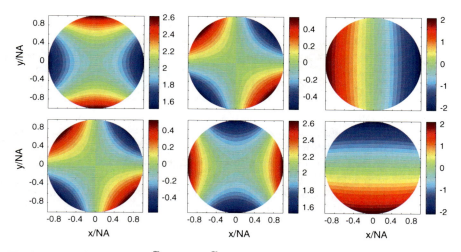

Fig. 8.3 Field distributions for $\widetilde{E}_x$ (*top*) and $\widetilde{E}_y$ (*bottom*) in the back focal plane of a collecting lens (NA = 0.966) arising from orthogonal electric dipoles orientated along the *x*, *y* and *z* coordinate axes respectively (*left* to *right*)

$$\widetilde{\mathbf{E}}_1(\theta_1, \phi_1) = \frac{1}{2\sqrt{\cos\theta_1}} \begin{pmatrix} (q_1 + q_2 \cos 2\phi_1) p_x + q_2 \sin 2\phi_1 p_y - q_3 \cos \phi_1 p_z \\ q_2 \sin 2\phi_1 p_x + (q_1 - q_2 \cos 2\phi_1) p_y - q_3 \sin \phi_1 p_z \\ 0 \end{pmatrix}, \quad (8.29)$$

where

$$q_1 = \cos\theta_1 + 1,$$
$$q_2 = \cos\theta_1 - 1,$$
$$q_3 = 2\sin\theta_1.$$

Equation (8.29) shows that the field in the back focal plane of the collector lens, can be considered as having contributions from three independent electric dipoles aligned with the Cartesian coordinate axes. Figure 8.3 shows the electric field components $\widetilde{E}_x$ (top row) and $\widetilde{E}_y$ (bottom row) in the back focal plane associated with each of these dipoles in turn (left to right).

Refocusing of the collected beam by a second low numerical aperture lens corresponds to a coherent, component-wise integration over the back focal plane.[4] Semi-analytic evaluation of the integrals can be achieved as has been discussed

---

[4] Strictly speaking the integration should be performed over the Gaussian reference sphere, however for a low numerical aperture lens the inter-component mixing caused by bending of rays of light is negligible and it is hence satisfactory to integrate over the back focal plane. The form of the image field of an electric dipole using two high numerical aperture lenses was considered earlier in Sect. 6.4.2.1 in which it was found the field on axis takes the form $\mathbf{E}_2 \propto (K_0^A p_x, K_0^A p_y, K_0^B p_z)$,

earlier, however the symmetry of the distributions in Fig. 8.3 immediately shows that at the geometric focus of the second lens, points in the integration plane will either cancel pairwise or superpose constructively. For example, for a dipole oriented parallel to the x-axis, the y component of the focused field must be zero since for each point (θ_1, ϕ_1) in the back focal plane there is a corresponding point of equal amplitude, yet out of phase by π at $(\theta_1, \phi_1 + \pi/2)$. Since the optical path length from each of these points to the geometric focus is equal they destructively interfere to give a null signal. However, for the x component, each point in the back focal plane is in phase (although of differing amplitudes) resulting in constructive interference at the detector. The converse is true for a dipole oriented parallel to the y axis, whilst for a purely longitudinal dipole there is destructive interference in *both* the x and y field components at the focus. The resulting on-axis focused field thus takes the form $\mathbf{E}_2 \propto (p_x, p_y, 0)$ in agreement with Sect. 6.4.2.1.

This result encapsulates the difficulty in determining the p_z component of a dipole, however given the arguments above it is evident that if the symmetry of the distributions in the back focal plane of the collector lens is broken, a significant p_z dependence can be introduced into the "image" of the dipole. Symmetry breaking of this type could be done, for example, by apodisation or phase modification of the beam; the latter being preferable in single molecule experiments, since phase masks do not reduce the optical throughput of the system. Imposition of a π phase delay to the collimated beam in the 1st and 4th quadrants with respect to the beam in the 2nd and 3rd quadrants, for example, gives rise to constructive interference in the E_x component when imaging a purely longitudinal dipole. Such a mask, placed in the back focal plane of the collector lens, is described by the amplitude transmittance function

$$T(x, y) = \begin{cases} -1 \text{ for } x \geq 0 \\ +1 \text{ for } x < 0 \end{cases}, \qquad (8.30)$$

and in its simplest form could be implemented using a glass block of appropriate thickness placed across one half of the beam as shown in Fig. 8.4a. Alternative implementations could however include use of a Pockels cell or a liquid crystal modulator. As such the optical setup shown in Fig. 8.5 and proposed by the author and colleagues in [7] provides a means to detect the full 3D orientation of an electric dipole in real time.

In the setup of Fig. 8.5 light collected from a dipole is incident into a beam splitter from which one of the output beams is further passed through a Wollaston prism (WP) which splits the field into its constituent $\widetilde{E}_x$ and $\widetilde{E}_y$ components. The other portion of light output from the beam splitter is passed through the phase mask described by Eq. (8.30) so as to break the symmetry in the beam profile. Finally the field in each arm is refocused onto point detectors (such as avalanche photodiodes) which respectively record the intensities D_1, D_2 and D_3. The use of point detectors ensures the detection process is field sensitive such that $D_1 = C_x |p_x|^2$, $D_2 = C_y |p_y|^2$ and

---

(Footnote 4 continued.)
where K_0^A and K_0^B are constants and $K_0^B / K_0^A \ll 1$. This arrangement will be neglected in this chapter however due to its practical difficulties.

8.3 Longitudinal Dipole Orientation and Field Mapping

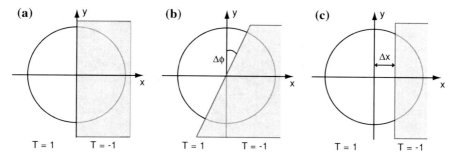

Fig. 8.4 **a** Proposed phase mask with **b** rotational misalignment or **c** translational misalignment

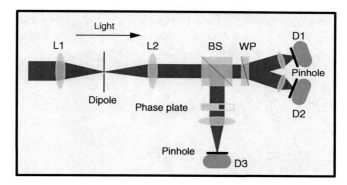

Fig. 8.5 Proposed optical setup for determination of the full three-dimensional orientation of an electric dipole. Notation is as follows: *L1* illuminating lens, *L2* collector lens, *BS* beam splitter, *WP* Wollaston prism, *D* detector

$D_3 = C_z |p_z|^2$. The effect of relaxing this assertion will be considered in the next section.

The constants C_x, C_y and C_z unfortunately need to be calibrated before measurements can be made, as can be performed using a known sample. It is however reasonable to assume that $C_x = C_y$ since the two detection arms are identical albeit for a 90° rotation in the state of polarisation. Furthermore, for orientational measurements it is only necessary to work with ratios of detector signals implying that only the value of the constant $C = C_x/C_z$ need be found as can be done using a gold bead or other similar point scatterer. Point scatterers of this type behave as free dipoles in which an effective electric dipole moment $\mathbf{p}_{\text{eff}}$ is induced proportional to the electric field vector of the illumination $\mathbf{E}_{\text{ill}}$ [25, 26]. Theoretically, the three detector signals will hence map the focused field distribution arising from the illumination beam (which can be calculated exactly using the Debye-Wolf integral) as the location of the scatterer is scanned in the object plane. It is then a simple matter to determine the constant of proportionality to determine C. Since C is dependent on the splitting ratio of the beam splitter it can be controlled to some extent, but does cause a trade-off of SNR between the different detector arms.

Point scatterers not only provide a means to calibrate the system of Fig. 8.5 but they also provide a means by which the full 3D electric field vector of an illuminating field can be found. The proposed optical setup hence provides a significant move towards full 3D polarimetry. In its current form there is an ambiguity as to which quadrant a measured dipole moment (or equivalently polarisation of the illuminating field) lies in, since only the magnitude of the three dipole components is measured. Introduction of additional arms in a similar fashion to a DOAP does however allow these ambiguities to be overcome, although the light incident on each detector is reduced hence having implications on the achievable measurement accuracy (see Sect. 5.2.4). Finally it is worth mentioning that if measuring a fixed dipole of moment **p**, in which the radiated field has strength proportional to $\mathbf{p} \cdot \mathbf{E}_{ill}$ it becomes important to match the excitation field to the specific detection needs (see Sect. 6.3.2 for an example involving measurement of transverse dipole orientation).

8.3.2 System Tolerances

Of importance in any practical implementation of the proposed detection scheme are the experimental tolerances of the phase mask. Departure from the ideal setup introduced above may arise in many guises, such as mask misalignments, attenuation and finite sized detectors. Each of these and a number of additional effects will be considered in an attempt to characterise the proposed system.

8.3.2.1 Mask Misalignments

Consider first misalignments of the phase mask, which in general can be treated as a combination of both a rotational and translational misalignment as depicted in Fig. 8.4. A rotational misalignment by an angle $\Delta\phi$ modifies the transmittance function, however assuming a symmetric pupil (which is normally the case) integration over the field does not yield any dependence on $\Delta\phi$. The detector signals are thus insensitive to pure rotational misalignments, a property which can again be attributed to the symmetrical nature of the back focal plane field distributions. Translational misalignments of the mask do however have a detrimental effect on the detector signals. A horizontal shift of, say, Δx modifies the transmittance function to

$$T(x, y) = \begin{cases} -1 & \text{for } x \geq \Delta x \\ +1 & \text{for } x < \Delta x \end{cases} \qquad (8.31)$$

which in general causes a mixing of the p_x and p_y signals into D_3, since destructive interference for these cases is no longer complete. To parameterise the extent of this mixing an extinction ratio is defined as the ratio of the detected intensity in D_3 when

8.3 Longitudinal Dipole Orientation and Field Mapping

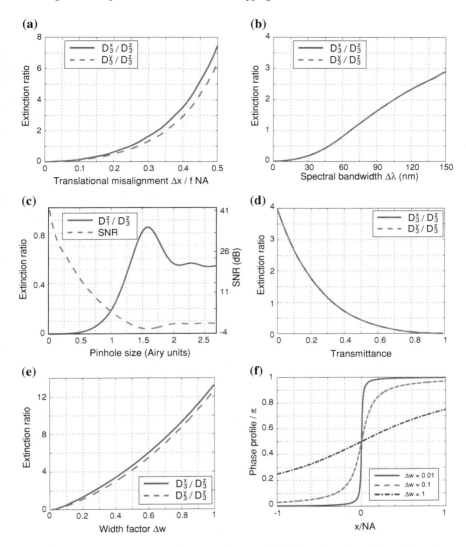

Fig. 8.6 Extinction ratios of the optical setup shown in Fig. 8.5 as a function of (**a**) mask misalignment, (**b**) spectral bandwidth, (**c**) detector aperture size, (**d**) mask attenuation and (**e**) width of phase mask transition as depicted in (**f**); **b** also shows the simulated SNR for different detector sizes

measuring a dipole parallel to the x (or y) axis, denoted D_3^x (and D_3^y), to that recorded for a z oriented dipole (D_3^z). Figure 8.6a shows the worsening of the extinction ratio as the mask is moved from the ideal position to being completely absent from the beam. Numerical simulations assumed a wavelength of 450 nm, numerical apertures of 0.95 and 0.01 for the collector and imaging lens respectively, and a beam splitting ratio of 2:1 such that each detection path receives equal power.

8.3.2.2 Finite Source Spectral Bandwidth

Hitherto discussion has been limited to monochromatic light of wavelength λ. This is however an unrealistic assumption if measuring the orientation of fluorescent molecules since they each emit a characteristic spectrum $S(\lambda)$ of wavelengths with an associated bandwidth $\Delta\lambda$. A non-zero bandwidth has consequences with regards to the glass slab used to impose the π phase delay since it can only be designed to operate perfectly at a single wavelength. The total intensity recorded by each detector can be calculated as an incoherent superposition of that measured for each wavelength individually such that

$$D_j^{\text{tot}} = \int_{\lambda_{\min}}^{\lambda_{\min}+\Delta\lambda} S(\lambda) D_j(\lambda; \lambda_0) d\lambda, \qquad (8.32)$$

where $D_j(\lambda; \lambda_0)$ is the detector reading ($j = 1, 2, 3$) for light of wavelength λ in the presence of a phase plate designed for operation at λ_0 and $\lambda_{\min}$ is the lower bound on the spectrum.

Results from numerical simulations assuming a Lorentzian spectrum profile,

$$S(\lambda) = \frac{2\Delta\lambda}{\Delta\lambda^2 + 4\pi^2(\lambda - \lambda_0)^2}, \qquad (8.33)$$

centered on the peak emission wavelength, $\lambda_0 = 455$ nm, of Pacific Blue [12] (a common fluorescent dye) are shown in Fig. 8.6b. Other spectral profiles give similar results. Neglecting chromatic aberrations that may occur in the focusing lens, no variation is seen in the extinction ratios for detectors D_1 and D_2 and thus only the extinction ratios for D_3 are shown. Inspection of this plot shows that the p_x and p_y signals mix into the D_3 signal and become double that of the p_z measurement for a bandwidth of approximately 100 nm. Although typical fluorophores have bandwidths of ~30 nm this mixing effect is likely to limit the performance of the setup. Use of achromatic waveplates, typically with a bandwidth of a few hundred nanometers, to implement the phase mask would however greatly improve the situation. The results of Fig. 8.6b thus represent the worst case performance.

8.3.2.3 Finite Sized Detectors

As a further consideration the previous assertion of point detectors is relaxed. In general D_1 and D_2 operate as confocal microscopes with finite sized pinholes illuminated with x and y polarised light respectively, an analysis of which can be found in [24]. Consequently only the modified detection path need be considered. Finite sized detectors record the integrated intensity of the focused light distribution

8.3 Longitudinal Dipole Orientation and Field Mapping

over their spatial extent, however as discussed in Sect. 6.4.2.1 the off-axis image of a dipole is dependent on all three dipole components [25]. For a dipole with moment $\mathbf{p} = (0, 0, 1)$ parasitic signals are thus introduced into detectors D_1 and D_2. Treating these signals as a noise source, a SNR can be defined as $10 \log_{10} \left(D_3^z / 2 D_1^z \right)$, the behaviour of which is shown in Fig. 8.6c (the factor of two is introduced since the parasitic signal is introduced to two detectors). The extinction ratio $D_1^z / D_3^z = D_2^z / D_3^z$ is also plotted for comparison. Typical values of the SNR in single molecule experiments are of the order of 15 dB (Fig. 8.1) and as such Fig. 8.6c shows that the effect of a finite detector size becomes a dominant factor once the radius of the detector is approximately half the size of the Airy disc. Detectors smaller than this are thus preferable. It is also found that the SNR arising when measuring a pure p_z dipole is worse than for measuring pure p_x and p_y dipoles and is hence the limiting case.

8.3.2.4 Mask Imperfections

Non-ideal characteristics of the phase mask are also likely to affect the performance of the optical setup and two such imperfections have thus also been modelled. The first of these is to introduce an attenuation into the beam to represent possible absorption by the mask (zero attenuation corresponds to complete transmittance), such that the transmission function is given by

$$T(x, y) = \begin{cases} -T_0 & \text{for } x \geq 0 \\ +1 & \text{for } x < 0 \end{cases} \tag{8.34}$$

where $0 \leq T_0 \leq 1$ is the transmittance of the phase mask. An imperfection of this nature, again destroys the complete destructive interference in the third arm when measuring pure p_x and p_y dipoles, giving rise to parasitic signals. Extinction ratios for these signals for differing degrees of attenuation were calculated and are shown in Fig. 8.6d.

Finally, the phase mask of Fig. 8.4 ideally requires a sharp discontinuity in the the imposed phase profile. Physical realisation of such a sharp phase discontinuity is unlikely and thus the effect of a continuous phase profile, defined by

$$\Delta \Phi = \arctan \left(\frac{1}{\Delta w} \frac{x}{f \text{NA}} \right) \tag{8.35}$$

and shown in Fig. 8.6f for different transition width factors Δw, was considered. The resulting extinction ratios are shown in Fig. 8.6e. From Fig. 8.6d, e it is seen that the setup is markedly more susceptible to imperfections which alter the phase of the back focal plane field distributions than to amplitude changes. In as far as the principle of operation of the proposed setup relies on interference, be it constructive or destructive, at the detector plane such an increased dependence would be expected.

8.4 Conclusions

Focus throughout this chapter has centered on single molecule studies, which is fast becoming a principle scientific tool in many areas of research and development. Two distinct problems have been considered centering on orientational measurements which are frequently employed in such single molecule studies. The first problem concerned noise statistics that will unavoidably be present in any optical measurements made. It was seen through consideration of the relative size of noise present in single molecule experiments that not only will fundamental quantisation noise be present, but other fluctuations that may be present in system can often be comparable in size. In particular, the issue of rotational motions of single molecules, be they discrete jumps or continuous angular diffusion, during the course of a measurement was considered and PDFs for the detector photoncount derived. Although results derived were valid for all timescales over which such rotations can occur a number of limiting cases were given. Furthermore, by study of the resulting fluctuations in the detector photoncount, regimes were identified under which different noise sources were seen to be dominant, an important aspect in terms of system design and noise analysis.

The second and separate issue addressed, was the determination of the longitudinal orientation of single molecules; something which has long eluded scientists. Symmetry considerations gave rise to the design of a simple optical system capable of measuring the full 3D orientation of single molecules. Not only could such a system open new opportunities and improve existing techniques in, say, biological research, but it also brings the possibility of 3D polarimetry to the fore. By exposing an additional degree of freedom of physical fields to experimental measurement additional informational gains may be possible in the future.

References

1. T. Basché, W.P. Ambrose, W.E. Moerner, Optical spectra and kinetics of single impurity molecules in a polymer: spectral diffusion and persistent spectral hole burning. J. Opt. Soc. Am. B **9**, 829–836 (1992)
2. R.E. Dale, S.C. Hopkins, Model-independent analysis of the orientation of fluorescent probes with restricted mobility in muscle fibers. Biophys. J. **76**, 1606–1618 (1999)
3. P. Debye, Polar molecules. Ph.D. Thesis, Dover, New York, 1945
4. L.A. Deschenes, D.A. van den Bout, Single-molecule studies of heterogeneous dynamics in polymer melts near the glass transition. Science **292**, 233, 255–258 (2001)
5. R.M. Dickson, D.J. Norris, W.E. Moerner, Simultaneous imaging of individual molecules aligned both parallel and perpendicular to the optic axis. Phys. Rev. Lett. **81**, 5322–5325 (1998)
6. W. Feller, *Probability Theory and its Applications* (Addison-Wesley, New York, 1950)
7. M.R. Foreman, C. Macías Romero, P. Török, Determination of the three dimensional orientation of single molecules. Opt. Lett. **33**, 1020–1022 (2008)
8. I.S. Gradshteyn, I.M. Ryzhik, *Table of Integrals, Series and Products* (Elsevier Academic Press, New York, 1980)

References

9. T. Ha, T. Enderle, D.S. Chemla, P.R. Selvin, S. Weiss, Single molecule dynamics studied by polarization modulation. Phys. Rev. Lett. **77**, 3979–3982 (1996)
10. T. Ha, T. Enderle, D.F. Ogletree, D.S. Chemla, P.R. Selvin, S. Weiss, Probing the interaction between two single molecules: fluorescence resonance energy transfer between a single donor and a single acceptor. Proc. Natl. Acad. Sci. U S A **93**, 6264–6268 (1996)
11. T. Ha, J. Glass, T. Enderle, D.S. Chemla, S. Weiss, Hindered rotational diffusion and rotational jumps of single molecules. Phys. Rev. Lett. **80**, 2093–2096 (1998)
12. Invitrogen, Fluorescence spectraviewer, http://www.invitrogen.com/site/us/en/home/support/Research-Tools/Fluorescence-SpectraViewer.html
13. K. Itô, *Introduction to Probability Theory* (Cambridge University Press, Cambridge, 1984)
14. T.M. Jovin, M. Bartholdi, W.L.C. Vaz, R.H. Austin, Rotational diffusion of biological macromolecules by time-resolved delayed luminescence (phosphorescence, fluorescence) anisotropy. Ann. N Y Acad. Sci. **366**, 176–196 (1981)
15. A. Leon-Garcia, *Probability and Random Processes for Electrical Engineering* (Addison-Wesley, New York, 1994)
16. H.P. Lu, L.Y. Xun, X.S. Xie, Single molecule enzymatic dynamics. Science **282**, 1877–1882 (1998)
17. W.E. Moerner, D.P. Fromm, Methods of single molecule fluorescence spectroscopy and microscopy. Rev. Sci. Instrum. **74**, 3597–3619 (2003)
18. I. Munro, I. Pecht, L. Stryer, Subnanosecond motions of Tryptophan residues in proteins. Proc. Natl. Acad. Sci. U S A **76**, 56–60 (1979)
19. D. Patra, I. Gregor, J. Enderlein, Image analysis of defocused single molecule images for three dimensional molecular orientation studies. J. Phys. Chem. A **108**, 6836 (2004)
20. G.H. Patterson, S.N. Knobel, W.D. Sharif, S.R. Kain, D.W. Piston, Use of the green fluorescent protein and its mutants in quantitative fluorescence microscopy. Biophys. J. **73**, 2782–2790 (1997)
21. D.J. Pikas, S.M. Kirkpatrick, E. Tewksbury, L.L. Brott, R.R. Naik, M.O. Stone, W.M. Dennis, Nonlinear saturation and lasing characteristics of green fluorescent protein. J. Phys. Chem. B **106**, 4831–4837 (2002)
22. B. Sick, B. Hecht, L. Novotny, Orientational imaging of single molecules by annular illumination. Phys. Rev. Lett. **85**, 4482–4485 (2000)
23. P. Török, Propagation of electromagnetic dipole waves through dielectric interfaces. Opt. Lett. **25**, 1463–1465 (2000)
24. P. Török, P.D. Higdon, T. Wilson, On the general properties of polarised light conventional and confocal microscopes. Opt. Commun. **148**, 300–315 (1998)
25. P. Török, P.D. Higdon, T. Wilson, Theory for confocal and conventional microscopes imaging small dielectric scatterers. Opt. Commun. **45**, 1681–1698 (1998)
26. H.C. van de Hulst, *Light Scattering by Small Particles* (Dover Publications, Dover, 1981)
27. P. Wahl, K. Tawada, J.C. Auchet, Study of tropomyosin labelled with a fluorescent probe by pulse fluorimetry in polarized light—interaction of that protein with troponin and actin. Eur. J. Biochem. **88**, 421–424 (1978)
28. D.M. Warshaw, E. Hayes, D. Gaffney, A.M. Lauzon, J.R. Wu, G. Kennedy, K. Trybus, S. Lowey, C. Berger, Myosin conformational states determined by single fluorophore polarization. Proc. Natl. Acad. Sci. U S A **95**, 8034–8039 (1998)
29. S. Weiss, Measuring conformational dynamics of biomolecules by single molecule fluorescence spectroscopy. Nat. Struct. Biol. **7**, 724 (2000)
30. K.D. Weston, L.S. Goldner, Orientation imaging and reorientation dynamics of single dye molecules. J. Phys. Chem. B **105**, 3453–3462 (2001)
31. J. Yguerabide, H.F. Epstein, L. Stryer, Segmental flexibility in an antibody molecule. J. Mol. Biol. **51**, 573–590 (1970)

Chapter 9
Conclusions

> *In light of knowledge attained, the happy achievement seems almost a matter of course, and any intelligent student can grasp it without too much trouble. But the years of anxious searching in the dark, with their intense longing, their alterations of confidence and exhaustion and the final emergence into the light—only those who have experienced it can understand it.*
> Albert Einstein

Central to this thesis has been the characterisation and exploitation of the opportunities afforded by the electromagnetic (i.e. vectorial) nature of light. To this end the work presented can be seen to follow one of three running themes: quantification of polarisation information; analytic formulations so as to simplify the propagation of electromagnetic waves; and development of specific polarisation based optical systems.

Characterising the informational limits inherent in polarisation based optical systems in essence reduces to considering the uncertainty present in any observations. Uncertainty can for example arise from a stochastic variation in the polarisation of light being measured, or from random noise perturbations to detector readings. Fisher information was deemed to be a suitable metric by which to measure the limits imposed by such sources of uncertainty as conveniently expressed by the CRLB. From the basis of statistical estimation theory, Stokes space was introduced and Fisher information used to define a polarisation resolution within this space, as is germane to Stokes polarimetry. Extension to a 16D Mueller space as appropriate for Mueller polarimetry is possible, indeed the definition is applicable in an appropriate Hilbert space associated with inference of any parameter vector, such as those found by means of a Lu-Chipman polar decomposition.

Informational limits in vectorial optics, at least within a classical framework, are found not to be absolute, but instead are fundamentally set by the mean number of information carrying photons. This is in stark contrast to many existing definitions, such as Rayleigh's criterion. Fisher information can be formulated within a quantum context, however the extent to which the conclusions drawn are affected, say by incorporation of the commutation relations on Stokes parameters, has yet to be fully determined. Further to the definition of polarisation resolution, alternative metrics, such as the number of degrees of freedom and the efficiency of observation were also defined as may be more suitable for multiplexed systems or experiments with low light levels.

Through maximisation of the polarisation resolution (or D-optimality) polarimetric systems can be optimised. Whilst, within the limited assumptions of existing design strategies, optimisation with regards to Fisher information was found to give consistent results, such as signal equalisation between multiple detectors and maximally distant measurement projection states, the technique developed within this thesis allows a more holistic approach to be adopted. For example the Fisher information formalism allows easy incorporation of both signal dependent noise models (e.g. Poisson noise arising from photon counting) and complex post-detection processing. Analytic results were given in this vein, with regards to the Mueller matrix decomposition of Lu and Chipman. Probabilistic a priori information can furthermore be easily included, allowing additional gains to be made. It is hence the author's belief that while existing optimisation practices are still invaluable, and frequently more straightforward to implement, careful consideration must be given by system designers as to whether further influences are at play and whether the additional gains then achievable using the Fisher information formalism are necessary. It is envisaged that D-optimality will thus be sought in scenarios in which high precision is needed, photon numbers are low, or a priori information is possessed, such as astronomy or communications.

Whilst standard polarimetry and the associated optimisation routines are sufficient when considering homogeneously polarised light or measurement at a point, in a polarisation imaging context the situation becomes unsatisfactory. Due to the mixing of the electric field components in high numerical aperture focusing high image fidelity is generally unachievable and an informatic approach is more suitable. Motivated by such considerations Fisher information was further employed to analyse inference problems in polarisation microscopy. Specifically electric dipole sources were considered and potential coherent and incoherent crosstalk between multiple dipoles, investigated in the context of orientational measurements. Although formulated in terms of electric dipoles, similar results can be found for magnetic dipoles. Critically it was shown that poor accuracy frequently results when parameter inference is based on zero readings, due to the increased sensitivity to unknown noise sources. Increased redundancy and averaging introduced in an imaging scenario can, however, help to compensate for such poor performance.

Prerequisite to the modelling and analysis of more advanced polarisation based imaging systems is the apparatus by which to describe the propagation of arbitrary electromagnetic waves in such systems. Whilst a Green's tensor formulation presents a suitable solution, it frequently requires significant computational resources, a situation which can be avoided by closer consideration of the problem. A new formulation, based on the scaled Debye-Wolf diffraction integral and generalised Jones calculus, was thus developed and presented for focusing of electromagnetic beams:

an integral part of any imaging system. Intentionally the treatment was kept as general as possible, allowing beams of arbitrary spatially inhomogeneous polarisation and coherence properties to be focused by systems of arbitrary numerical aperture and Fresnel number, something which to date has not been possible. Simplifications were also presented by employing a coherent mode expansion. Owing to continuing debate in the literature, both scalar- and vector based coherent mode representations were used. In so doing it was shown that computational gains and improved mathematical tractability are achievable with an assumption of harmonic angular variation, a somewhat less restrictive imposition than the customary assumption of rotational symmetry. Ultimately the developed theory is suitable for application in a wider range of problems beyond those examined in this thesis, however extension to informational considerations when imaging non-dipolar sources still requires further investigation.

Restricting to fully coherent fields an eigenfunction expansion was devised to further complement the description of focusing. That said, a single coherent mode, being by construction spatially coherent, can be represented by means of the eigenfunction expansion and hence the field distribution in the focal plane easily calculated.

Whilst the expansion was seen to possess a number of useful computational properties, particular benefit can be drawn from the inherent structure of an eigen representation and its suitability for solution of inverse problems. Unfortunately the constraints imposed by Maxwell's equations are not automatically fulfilled by the developed series since each field component is expanded separately, thus presenting complications when attempting to solve such problems. Particularly, in the specification of an inverse focusing problem one (or more) field components must be left unconstrained, which was demonstrated to hinder significant resolution gains in the focused intensity distribution, as may be desirable for increasing the information available in polarisation imaging or optical data storage. Resolution of this drawback requires solution of an integral equation with matrix-valued kernel (hence yielding vector-valued basis functions), however analytic results have not yet been forthcoming, despite the author's efforts, and hence remains as future work. Nevertheless, given the fast convergence of the developed series expansion and hence the relatively few number of associated coefficients, numerical optimisation routines can still be utilised without difficulty as has been illustrated using an EDF example. Numerical optimisation can be further guided by the insights gained by the nature of the basis functions. The expansion is furthermore still of significant use in applications in which the response of a system to each field component differs or needs to be tailored.

Finally, in the closing chapters of this thesis, two specific examples of how polarisation can be utilised in current problems in optics was explored. Encoding data into the orientation of asymmetrical pits was numerically modelled from which it was determined that by measurement of the state of polarisation of the scattered light it is possible to simply infer the orientation of a pit. It was found that elongated data pits act as near ideal optical diattenuators due to the differing phase shifts introduced into the scattered x and y field components. Polarisation multiplexed optical data storage was thus seen to be a potential candidate for increasing the data capacity of optical media. This novel multiplexed optical data storage solution was further investigated by considering how the diattenuation of a pit could be maximised, consequently maximising the storage capabilities of a MODS system. Quantitative discrepancies with recent experimental results were found and discussed, albeit qualitative agreement was seen. The exact sources of these differences still need to be resolved, however principally are believed to originate from the assumption of perfect conductivity.

The second, and final, example examined exploiting polarisation properties of light in single molecule studies. Given the typically low SNR ratios in single molecule experiments and the consequent photon counting required, it was seen that stochastic variations in the measured signal arising from both the discrete nature of light and other potential source fluctuations can play equally important roles in measurement accuracy. Again following the theme of orientational measurements, potential reorientational changes of the single molecule (and the ensuing change in polarisation of radiated light) during the course of a time integrated measurement was closely investigated, with both discrete and continuous orientational changes analysed. The associated PDF of the measured intensity was derived under general rates of rotational motion and a closed form given for the limiting cases of fast and slow orientational change. From such considerations appropriate regimes of dominant noise were identified, which will be valuable in determining the correct estimation protocols and noise analysis in single molecule experiments.

PDFs for both 2D and 3D rotational motions were derived, however determination of the longitudinal orientation of a single molecule, has, to date, been limited. The possibility of making such longitudinal measurements however would offer significant promise for full 3D polarimetry and biological research. Via consideration of the symmetry and resulting constructive and destructive interference, a means of doing so was identified and demonstrated rigorously. Experimental verification of the results are still pending, however it was shown mathematically that by breaking the symmetry in the back pupil plane of a simple $4f$ imaging system it is possible to modify the on-axis polarisation in the image plane so as to depend on the longitudinal dipole moment of the molecule. Accordingly a new three-arm DOAP arrangement, employing a half beam phase mask, was proposed so as to measure the full 3D orientation of a single molecule. Experimental tolerances of the setup were also investigated. The system performance, whilst insensitive to pure rotational misalignments of the phase mask, was found to deteriorate rapidly for translational misalignments of more than approximately 10% of the pupil width, well within practical limits. Furthermore given the typical bandwidth of emission spectra of single molecules only minor system degradation can be expected. Whilst other influences were also investigated it was primarily found that good system performance could be expected over a wide range of experimental conditions, however, since the proposed system is dependent on tailoring the interference properties in the image plane, it is more sensitive to phase perturbations, e.g. lens aberrations, than to amplitude variations.

9 Conclusions

In closing it is perhaps appropriate to paraphrase Ernest Rutherford in noting that there are two kinds of problems: impossible and trivial. All problems are impossible until you solve them, at which point they become trivial. Perhaps the most interesting things to come from this thesis are thus not necessarily the results and tools developed, although it is certainly hoped that these prove useful and interesting to others, but instead the questions that still remain unanswered and the new problems posed. Such is the nature of science and the author would not wish it any other way.

Appendix A
Some Information Theoretic Proofs

A number of results have been quoted from information and estimation theory throughout the main body of this text. In this appendix, proofs for two such results are given. Firstly, a derivation of the multivariate CRLB when estimating complex parameters in the presence of bias is presented in Sect. A.1, since although the author is confident this result is known, a suitable reference containing a proof for such a general scenario appeared lacking. Secondly, in Sect. A.2 a derivation for the multivariate BCRLB is presented since again an explicit proof is difficult to find in the literature.

A.1 Multivariate Cramér-Rao Lower Bound for Biased Estimation

A bound is sought on the covariance matrix, $\mathbb{K}_\mathbf{w}$, of an estimate $\hat{\mathbf{w}}$ of the complex parameter vector $\mathbf{w}$. As such consider constructing a vector

$$\mathbf{v} = \begin{pmatrix} \hat{\mathbf{w}} - E_\mathbf{X}[\hat{\mathbf{w}}] \\ \mathbf{s_w} \end{pmatrix}, \tag{A.1}$$

where $\mathbf{s_w} = (\partial \ln f_\mathbf{X}(\mathbf{x}|\mathbf{w})/\partial \mathbf{w})^\dagger$ is the score vector and $f_\mathbf{X}(\mathbf{x}|\mathbf{w})$ is the likelihood function for the noisy observations $\mathbf{x}$ as parameterised by $\mathbf{w}$. Forming the correlation matrix

$$E_\mathbf{X}[\mathbf{v}\mathbf{v}^\dagger] = E_\mathbf{X} \begin{bmatrix} (\hat{\mathbf{w}} - E_\mathbf{X}[\hat{\mathbf{w}}])(\hat{\mathbf{w}} - E_\mathbf{X}[\hat{\mathbf{w}}])^\dagger & (\hat{\mathbf{w}} - E_\mathbf{X}[\hat{\mathbf{w}}])\mathbf{s_w}^\dagger \\ \mathbf{s_w}(\hat{\mathbf{w}} - E_\mathbf{X}[\hat{\mathbf{w}}])^\dagger & \mathbf{s_w}\mathbf{s_w}^\dagger \end{bmatrix},$$
$$= \begin{pmatrix} \mathbb{K}_\mathbf{w} & (\mathbb{I} + \frac{\partial \mathbf{b_w}}{\partial \mathbf{w}}) \\ (\mathbb{I} + \frac{\partial \mathbf{b_w}}{\partial \mathbf{w}})^\dagger & \mathbb{J}_\mathbf{w} \end{pmatrix}, \tag{A.2}$$

where the definition of the bias $\mathbf{b_w}$ of an estimator, $\mathbf{b_w} = E_\mathbf{X}[\hat{\mathbf{w}}] - \mathbf{w}$ has been used and the FIM has been defined as

$$\mathbb{J}_\mathbf{w} = E_\mathbf{X}\left[\left(\frac{\partial \ln f_\mathbf{X}(\mathbf{x}|\mathbf{w})}{\partial \mathbf{w}}\right)^\dagger \frac{\partial \ln f_\mathbf{X}(\mathbf{x}|\mathbf{w})}{\partial \mathbf{w}}\right]. \tag{A.3}$$

By virtue of being a correlation matrix, $E_\mathbf{X}[\mathbf{v}\mathbf{v}^\dagger]$ is positive semi-definite implying $\alpha^\dagger E_\mathbf{X}[\mathbf{v}\mathbf{v}^\dagger]\alpha \geq 0$. Letting

$$\alpha = \begin{pmatrix} \beta \\ -\mathbb{J}_\mathbf{w}^{-1}\left(\mathbb{I} + \frac{\partial \mathbf{b_w}}{\partial \mathbf{w}}\right)^\dagger \beta \end{pmatrix} \tag{A.4}$$

and evaluating the matrix multiplication yields

$$\beta^\dagger\left[\mathbb{K}_\mathbf{w} - \left(\mathbb{I} + \frac{\partial \mathbf{b_w}}{\partial \mathbf{w}}\right)\mathbb{J}_\mathbf{w}^{-1}\left(\mathbb{I} + \frac{\partial \mathbf{b_w}}{\partial \mathbf{w}}\right)^\dagger\right]\beta \geq 0. \tag{A.5}$$

Equation (A.5) then in turn indicates that the matrix

$$\mathbb{K}_\mathbf{w} - \left(\mathbb{I} + \frac{\partial \mathbf{b_w}}{\partial \mathbf{w}}\right)\mathbb{J}_\mathbf{w}^{-1}\left(\mathbb{I} + \frac{\partial \mathbf{b_w}}{\partial \mathbf{w}}\right)^\dagger \tag{A.6}$$

is positive semi-definite, or equivalently

$$\mathbb{K}_\mathbf{w} \geq \left(\mathbb{I} + \frac{\partial \mathbf{b_w}}{\partial \mathbf{w}}\right)\mathbb{J}_\mathbf{w}^{-1}\left(\mathbb{I} + \frac{\partial \mathbf{b_w}}{\partial \mathbf{w}}\right)^\dagger. \tag{A.7}$$

Equation (A.7) represents the CRLB for estimation of complex parameters in the presence of bias as quoted in Eq. (3.27).

A.2 Multivariate Bayesian Cramér-Rao Lower Bound

When estimating a deterministic parameter vector $\mathbf{w}$ from the random data vector $\mathbf{x}$, the FIM is defined as

$$\mathbb{J}_\mathbf{w} = E_\mathbf{X}\left[\left(\frac{\partial \ln f_\mathbf{X}(\mathbf{x}|\mathbf{w})}{\partial \mathbf{w}}\right)^\dagger \frac{\partial \ln f_\mathbf{X}(\mathbf{x}|\mathbf{w})}{\partial \mathbf{w}}\right]. \tag{A.8}$$

If however $\mathbf{w}$ can vary, this definition becomes unsatisfactory since it does not account for any *a priori* knowledge about the random nature of the parameter that may be possessed and which can be used to improve the precision of any estimate

Appendix A: Some Information Theoretic Proofs

of $\mathbf{w}$. Instead it is more appropriate to define the FIM in terms of the joint PDF of $\mathbf{X}$ and $\mathbf{W}$, namely $f_{\mathbf{X},\mathbf{W}}(\mathbf{x},\mathbf{w}) = f_{\mathbf{X}}(\mathbf{x}|\mathbf{w})f_{\mathbf{W}}(\mathbf{w})$. Taking the logarithm gives

$$L(\mathbf{x},\mathbf{w}) = \ln f_{\mathbf{X}}(\mathbf{x}|\mathbf{w}) + \ln f_{\mathbf{W}}(\mathbf{w}), \tag{A.9}$$

so that the modified FIM is defined by

$$\mathbb{J}_{\mathbf{w}} = \sum_{i=1}^{2}\sum_{j=1}^{2} E_{\mathbf{X},\mathbf{W}}\left[\frac{\partial L_i}{\partial \mathbf{w}}^{\dagger} \frac{\partial L_j}{\partial \mathbf{w}}\right],$$
$$= \mathbb{J}_1 + \mathbb{J}_2 + \mathbb{J}_3 + \mathbb{J}_4, \tag{A.10}$$

where $L_1(\mathbf{x},\mathbf{w}) = \ln f_{\mathbf{X},\mathbf{W}}(\mathbf{x}|\mathbf{w})$ and $L_2(\mathbf{w}) = \ln f_{\mathbf{W}}(\mathbf{w})$, and expectations are now with respect to both $\mathbf{X}$ and $\mathbf{W}$. Considering each of these terms in turn gives

$$\mathbb{J}_1 = \iint \frac{\partial L_1}{\partial \mathbf{w}}^{\dagger} \frac{\partial L_1}{\partial \mathbf{w}} f_{\mathbf{X},\mathbf{W}}(\mathbf{x},\mathbf{w}) d\mathbf{x} d\mathbf{w},$$
$$= \int \left[\int \frac{\partial L_1}{\partial \mathbf{w}}^{\dagger} \frac{\partial L_1}{\partial \mathbf{w}} f_{\mathbf{X}}(\mathbf{x}|\mathbf{w}) d\mathbf{x}\right] f_{\mathbf{W}}(\mathbf{w}) d\mathbf{w}, \tag{A.11}$$
$$= E_{\mathbf{W}}[\mathbb{J}_{\mathbf{w}}^r],$$

where $E_{\mathbf{W}}[\ldots]$ denotes the expectation with respect to $\mathbf{W}$ only and $\mathbb{J}_{\mathbf{w}}^r$ is the Fisher information matrix as defined by Eq. (A.8).

Adopting a similar treatment of $\mathbb{J}_2$ gives

$$\mathbb{J}_2 = \iint \frac{\partial L_1}{\partial \mathbf{w}}^{\dagger} \frac{\partial L_2}{\partial \mathbf{w}} f_{\mathbf{X},\mathbf{W}}(\mathbf{x},\mathbf{w}) d\mathbf{x} d\mathbf{w},$$
$$= \iint \frac{f_{\mathbf{X},\mathbf{W}}(\mathbf{x},\mathbf{w})}{f_{\mathbf{X}}(\mathbf{x}|\mathbf{w})f_{\mathbf{W}}(\mathbf{w})} \frac{\partial f_{\mathbf{X}}(\mathbf{x}|\mathbf{w})}{\partial \mathbf{w}}^{\dagger} \frac{\partial f_{\mathbf{W}}(\mathbf{w})}{\partial \mathbf{w}} d\mathbf{x} d\mathbf{w}, \tag{A.12}$$
$$= \int \left[\frac{\partial}{\partial \mathbf{w}} \int f_{\mathbf{X}}(\mathbf{x}|\mathbf{w}) d\mathbf{x}\right]^{\dagger} \frac{\partial f_{\mathbf{W}}(\mathbf{w})}{\partial \mathbf{w}} d\mathbf{w},$$
$$= \mathbb{O},$$

where $\mathbb{O}$ is a matrix of zeros, as follows from $\int f_{\mathbf{X}}(\mathbf{x}|\mathbf{w})d\mathbf{x} = 1$. A similar result follows for $\mathbb{J}_3$. Finally consider

$$\mathbb{J}_4 = \iint \frac{\partial L_2}{\partial \mathbf{w}}^{\dagger} \frac{\partial L_2}{\partial \mathbf{w}} f_{\mathbf{X},\mathbf{W}}(\mathbf{x},\mathbf{w}) d\mathbf{x} d\mathbf{w},$$
$$= \int \frac{\partial L_2}{\partial \mathbf{w}}^{\dagger} \frac{\partial L_2}{\partial \mathbf{w}} f_{\mathbf{W}}(\mathbf{w}) d\mathbf{w}, \tag{A.13}$$
$$= \mathbb{J}_{\mathbf{w}}^{ap}.$$

Combining these results finally yields

$$\mathbb{J}_\mathbf{w} = E_\mathbf{W}\left[\mathbb{J}_\mathbf{w}^r\right] + \mathbb{J}_\mathbf{w}^{ap}. \tag{A.14}$$

It is thus evident that the FIM, when trying to estimate a random parameter $\mathbf{W}$, is given by the average of the Fisher information for a deterministic $\mathbf{w}$, with respect to $\mathbf{W}$, plus an additional term arising from *a priori* knowledge of the random behaviour of $\mathbf{W}$.

As a final point of interest, note that if $f_\mathbf{W}(\mathbf{w})$ is uniform then

$$\frac{\partial \ln f_\mathbf{W}(\mathbf{w})}{\partial \mathbf{w}} = \mathbf{0}, \tag{A.15}$$

such that $\mathbb{J}_\mathbf{w}^{ap} = \mathbb{O}$ and $\mathbb{J}_\mathbf{w} = E_\mathbf{W}\left[\mathbb{J}_\mathbf{w}^r\right]$. Furthermore if $\mathbb{J}_\mathbf{w}^r$ is *not* dependent on $\mathbf{w}$ then $\mathbb{J}_\mathbf{w} = \mathbb{J}_\mathbf{w}^r$.

Appendix B
Special Functions

B.1 Generalised Prolate Spheroidal Functions

In this section various properties of generalised prolate spheroidal functions are introduced and discussed. Although a number of different mathematical properties are considered, derivations are omitted for brevity. Reference is however made to the works of Slepian, Landau and Pollock [6, 7, 10, 12] where a full analysis can be found. A summary of the key properties is provided in Table B.1.

B.1.1 Space-Bandwidth Product

All optical devices are incapable of perfectly transmitting signals with arbitrarily high frequency content, but instead possess transfer functions which extend over a finite range. The resulting transmitted signal thus has a finite bandwidth denoted Ω. A bandlimited function cannot in itself also be space limited due to the uncertainty principle, however it is possible to define a region of spatial extent r_0 outside of which the function is negligible or of little interest. The product $c = r_0 \Omega$ is then called the space-bandwidth product and is often used as a measure of the optical performance of a system [8, 9].

The space-bandwidth product is important when discussing prolate spheroidal functions since they are bandlimited functions whose form and behaviour is dependent upon the parameter c. This explicit dependence is however occasionally dropped in this work for clarity with the understanding that the dependence still remains.

Table B.1 Summary of the properties of the spheroidal prolate functions with space-bandwidth product $c = r_0\Omega$, for $N \geq 0$ and $n \geq 0$

Finite Hankel transform (definition)	$\int_0^{r_0} \Phi_{N,n}(r) J_N(\omega r) r\, dr = (-1)^n \left(\frac{r_0}{\Omega}\right) \lambda_{N,n}^{1/2} \Phi_{N,n}\left(\frac{\omega r_0}{\Omega}\right)$
Infinite Hankel transform	$\int_0^\infty \Phi_{N,n}(r) J_N(\omega r) r\, dr = \begin{cases} (-1)^n \left(\frac{r_0}{\Omega}\right) \lambda_{N,n}^{1/2} \Phi_{N,n}\left(\frac{\omega r_0}{\Omega}\right), & \omega \leq \Omega \\ 0, & \omega > \Omega \end{cases}$
Orthogonality (finite spatial domain)	$\int_0^{r_0} \Phi_{N,n}(r) \Phi_{N,m}(r) r\, dr = \lambda_{N,n} \delta_{nm}$
Orthogonality (infinite spatial domain)	$\int_0^\infty \Phi_{N,n}(r) \Phi_{N,m}(r) r\, dr = \delta_{nm}$
Inverse finite Hankel transform	$\int_0^\Omega \Phi_{N,n}\left(\frac{\omega r_0}{\Omega}\right) J_N(\omega r) \omega\, d\omega = (-1)^n \left(\frac{\Omega}{r_0}\right) \lambda_{N,n}^{1/2} \Phi_{N,n}(r)$
Inverse infinite Hankel transform	$\int_0^\Omega \Phi_{N,n}\left(\frac{\omega r_0}{\Omega}\right) J_N(\omega r) \omega\, d\omega = \begin{cases} (-1)^n \left(\frac{\Omega}{r_0}\right) \lambda_{N,n}^{1/2} \Phi_{N,n}(r), & r \leq r_0 \\ 0, & r > r_0 \end{cases}$
Orthogonality (finite frequency domain)	$\int_0^\Omega \Phi_{N,n}\left(\frac{\omega r_0}{\Omega}\right) \Phi_{N,m}\left(\frac{\omega r_0}{\Omega}\right) \omega\, d\omega = \left(\frac{\Omega}{r_0}\right)^2 \lambda_{N,n} \delta_{nm}$
Orthogonality (infinite frequency domain)	$\int_0^\infty \Phi_{N,n}\left(\frac{\omega r_0}{\Omega}\right) \Phi_{N,m}\left(\frac{\omega r_0}{\Omega}\right) \omega\, d\omega = \left(\frac{\Omega}{r_0}\right)^2 \delta_{nm}$

B.1.2 Eigenfunctions of the Two Dimensional Finite Fourier Integral

It can be shown [10] that the eigenfunctions of the *finite* two dimensional Fourier transform over a circular domain can be written in the form

$$\psi_{N,n}(c, r, \theta) = \Phi_{N,n}(c, r) \frac{\cos N\theta}{\sin N\theta}, \quad N = 0, 1, 2, \ldots, \quad n = 0, 1, 2, \ldots, \quad (B.1)$$

where $\Phi_{N,n}(c, r)$, known as the circular prolate spheroidal functions, are the eigenfunctions of the Nth order finite Hankel transform. The defining relation for these functions can thus be expressed

$$\int_0^{r_0} J_N(\omega r) \Phi_{N,n}(c, r) r\, dr = (-1)^n \left(\frac{r_0}{\Omega}\right) \lambda_{N,n}^{1/2} \Phi_{N,n}\left(c, \frac{\omega r_0}{\Omega}\right), \quad (B.2)$$

where $J_N(\cdots)$ is the Nth order Bessel function of the first kind, ω and r are conjugate coordinates and $\lambda_{N,n}$ are the circular prolate spheroidal eigenvalues. Figures B.1 and B.2 show the behaviour of the eigenvalues and eigenfunctions respectively which are further discussed in Sect. B.1.4

It should be noted that the circular prolate functions $\Phi_{N,n}$ used here are scaled versions of those developed by Slepian, denoted $\varphi_{N,n}$, such that

$$\Phi_{N,n}(c, r) = \left(\frac{\lambda_{N,n}}{rr_0}\right)^{1/2} \varphi_{N,n}\left(c, \frac{r}{r_0}\right). \quad (B.3)$$

Appendix B: Special Functions

Fig. B.1 Circular prolate spheroidal eigenvalues for different orders (N and n) and space bandwidth products c

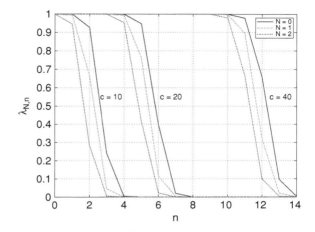

$\varphi_{N,n}$ are also the solutions to the wave equation when expressed in a prolate spheroidal coordinate system.

B.1.3 Orthogonality and Completeness of the Generalised Prolate Spheroidal Functions

Heurtley has shown [4] that the functions satisfying the integral Eq. (B.2) are orthogonal and complete over the finite region $0 \leq r \leq r_0$, i.e.

$$\int_0^{r_0} \Phi_{N,n}(c,r)\Phi_{N,m}(c,r) r\, dr = \lambda_{N,n}\delta_{nm} \tag{B.4}$$

and

$$\sum_{n=0}^{\infty} \lambda_{N,n}^{-1}\Phi_{N,n}(c,r)\Phi_{N,n}(c,r') = \frac{\delta(r-r')}{r} \quad \text{for} \quad 0 \leq r, r' \leq r_0, \tag{B.5}$$

where δ_{nm} is the Kronecker delta and $\delta(r-r')$ is the Dirac delta function centered on $r = r'$. Furthermore the prolate functions possess the unique property that they are also orthogonal and complete on the infinite interval $0 \leq r \leq \infty$.

Noting that harmonic exponentials are also complete and orthogonal it is possible to expand any two dimensional bandlimited function in terms of generalised prolate functions

$$f(r,\phi) = \sum_{N=-\infty}^{\infty} \sum_{n=0}^{\infty} A_{N,n}\Phi_{|N|,n}(c,r)\exp(iN\phi), \tag{B.6}$$

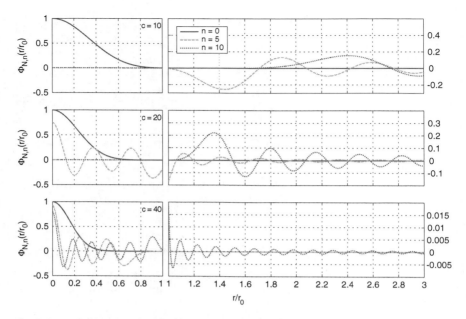

Fig. B.2 Circular prolate spheroidal functions for different orders n for $N = 0$ and for different space bandwidth products c. Prolate functions plotted have been normalised so that $\Phi_{0,0}(0) = 1$. Note the different vertical scales between plots

where it has been elected to write $\psi_{N,n}$ in terms of exponentials as opposed to the trigonometric functions of Eq. (B.1) for mathematical convenience.[1] The coefficients $A_{N,n}$ can be calculated using the orthogonality property whereby[2]

$$A_{N,n} = \frac{1}{2\pi \lambda_{|N|,n}} \int_0^{2\pi} \int_0^{r_0} f(r, \phi) \Phi_{|N|,n}(c, r) \exp(-iN\phi) r \, dr \, d\phi. \tag{B.7}$$

---

[1] Use of exponential angular terms is in contrast to the sinusoidal basis used in Slepian's original work. Since a linear combination of eigenfunctions of an operator is not necessarily also an eigenfunction it is important to verify the eigenequation is still satisfied. This is fortunately easily done by substitution of the basis functions $\Phi_{|N|,n}(c, r) \exp(iN\phi)$ into a finite 2D Fourier transform.

[2] The observant reader will notice that the circular prolate spheroidal eigenvalues are now written $\lambda_{|N|,n}$ as opposed to $\lambda_{N,n}$. The switch of index from N to $|N|$ is required due to the shift from a sinusoidal to an exponential angular term.

B.1.4 Energy Concentration Property of Generalised Prolate Spheroidal Functions

Given a bandlimited function the question may be asked as to how concentrated the function can be in the spatial domain in terms of its energy distribution. This is of particular interest in optics, for example when trying to improve the resolution in an imaging system or to extend the depth of field where large sidelobe structures are undesirable. Traditionally the encircled energy of a function $f(r, \phi)$ within a circular region of radius r_0 is defined as

$$I_{enc} = \int_0^{2\pi} \int_0^{r_0} |f(r,\phi)|^2 r dr d\phi \bigg/ \int_0^{2\pi} \int_0^{\infty} |f(r,\phi)|^2 r dr d\phi. \quad (B.8)$$

Expansion of $f(r, \phi)$ by using Eq. (B.6) and subsequent substitution into Eq. (B.8) gives

$$I_{enc} = \sum_{N=-\infty}^{\infty} \sum_{n=0}^{\infty} |A_{N,n}|^2 \lambda_{|N|,n} \bigg/ \sum_{N=-\infty}^{\infty} \sum_{n=0}^{\infty} |A_{N,n}|^2, \quad (B.9)$$

where the orthogonality condition (B.4) and the analogous equation for the infinite interval (see for example [3] or Table B.1) have also been used. From Fig. B.1 it can be seen that the eigenvalues lie in the range $0 \leq \lambda_{|N|,n} \leq 1$ and monotonically decrease with $|N|$ and n and as such the encircled energy takes its maximum value of $I_{enc}^{max} = \lambda_{0,0}$ when

$$f(r, \phi) = A_{0,0} \Phi_{0,0}(c, r). \quad (B.10)$$

More generally the eigenvalue $\lambda_{|N|,n}$ is a measure of the fraction of energy contained within the circular region defined by $0 \leq r \leq r_0$ and $0 \leq \phi < 2\pi$ [11]. This feature can be seen in Figs. B.1 and B.2. Considering, for example, first the $c = 10$ case it is noted that the eigenvalues drop off rapidly at $n \sim 3$. As such when the $n = 0$ order is plotted in Fig. B.2 it is non-zero when $r \leq 0$ and essentially (although not precisely) zero outside. Higher order modes, $n = 5$ and 10, however display the converse behaviour.

For the $c = 20$ case the eigenvalues remain close to unity up to higher orders and instead decrease at $n \sim 6$. When plotted the $n = 0$ mode displays the same properties as before, but now the $n = 5$ mode shows contributions for all values of r considered. With an eigenvalue of 8.46×10^{-9} the $n = 10$ order again contains negligible energy within the central region.

Finally considering the $c = 40$ case the eigenvalues do not fall off until $n \sim 13$ meaning the plotted orders have only a small contribution for $r \geq r_0$.

B.2 Electromagnetic Plane Waves

Much reference has been given to electromagnetic plane waves during the course of this thesis. In this section plane waves are derived as a solution of Maxwell's equations. The results given here will also be formulated in such a manner to allow the field modes supported in a rectangular waveguide to be derived in Sect. B.3.

Consider Maxwell's equations as given in Eqs. (4.1)–(4.4) and the wave equations that follow (c.f. Eq. (4.8))

$$\nabla^2 \mathbf{E} + k^2 \mathbf{E} = \mathbf{0}, \quad \nabla^2 \mathbf{H} + k^2 \mathbf{H} = \mathbf{0}, \tag{B.11}$$

where $k = \sqrt{\epsilon\mu}\omega$ is the wavenumber in the medium. An $\exp(-i\omega t)$ time dependence has been assumed throughout this work. These vector wave equations imply that each component of the electric and magnetic field must satisfy the Helmholtz equation individually. Since specialisation to a system with rectangular symmetry is later made, the Helmholtz equation is considered in Cartesian coordinates and solved using the technique of separation of variables [1]. The Helmholtz equation can then be written

$$\frac{\partial^2 U_j}{\partial x^2} + \frac{\partial^2 U_j}{\partial y^2} + \frac{\partial^2 U_j}{\partial z^2} + k^2 U_j = 0, \tag{B.12}$$

where $U_j(x, y, z) = X_j(x) Y_j(y) Z_j(z)$ represents a single component of either the electric or magnetic field and the separation of U_j has been given explicitly. Substitution of U_j in its separated form gives

$$\frac{1}{X_j}\frac{\partial^2 X_j}{\partial x^2} + \frac{1}{Y_j}\frac{\partial^2 Y_j}{\partial y^2} + \frac{1}{Z_j}\frac{\partial^2 Z_j}{\partial z^2} = -(k_x^2 + k_y^2 + k_z^2), \tag{B.13}$$

where $k^2 = k_x^2 + k_y^2 + k_z^2$. Since k_x, k_y and k_z are constants, Eq. (B.13) can be split into three uncoupled differential equations

$$\frac{d^2 X_j}{dx^2} = -k_x^2 X_j, \quad \frac{d^2 Y_j}{dy^2} = -k_y^2 Y_j, \quad \frac{d^2 Z_j}{dz^2} = -k_z^2 Z_j, \tag{B.14}$$

which have the general solutions

$$X_j(x) = A_j^+ \exp(ik_x x) + A_j^- \exp(-ik_x x), \tag{B.15a}$$

$$Y_j(y) = B_j^+ \exp(ik_y y) + B_j^- \exp(-ik_y y), \tag{B.15b}$$

$$Z_j(z) = C_j^+ \exp(ik_z z) + C_j^- \exp(-ik_z z), \tag{B.15c}$$

$A_j^\pm, B_j^\pm$ and $C_j^\pm$ are constants of integration set by boundary conditions, an example of which is discussed in Sect. B.3. In free space however no further constraints apply and hence Eqs. (B.15) can be combined to give

Appendix B: Special Functions

$$\mathbf{E} = \mathbf{E}^+ \exp(i\mathbf{k} \cdot \mathbf{r}) + \mathbf{E}^- \exp(-i\mathbf{k} \cdot \mathbf{r}). \tag{B.16}$$

Equation (B.16) represents the summation of two counter-propagating plane waves (propagating in directions defined by $\pm\mathbf{k}$).

B.2.1 s and p Polarised Plane Waves

Further to Eq. (B.16), a plane wave is frequently decomposed into two distinct polarisation components. Although the choice of decomposition is not unique, a decomposition into plane waves possessing a zero longitudinal electric or magnetic field component, so called *s*- and *p*-polarised waves, can prove expedient when solving plane interface problems, such as that considered in Chap. 7. Upon such a decomposition Eq. (B.16) can be written

$$\mathbf{E} = \sum_{\eta=\pm} \sum_{\nu=s,p} a^{\eta,\nu} \mathbf{E}^\nu(\mathbf{k}) \exp(i\eta \mathbf{k} \cdot \mathbf{r}), \tag{B.17}$$

or analogously for the magnetic field

$$\mathbf{H} = \sum_{\eta=\pm} \sum_{\nu=s,p} a^{\eta,\nu} \mathbf{H}^\nu(\mathbf{k}) \exp(i\eta \mathbf{k} \cdot \mathbf{r}), \tag{B.18}$$

where

$$\mathbf{E}^s(\mathbf{k}) = \begin{cases} \dfrac{\omega\mu}{2\pi\sqrt{k_x^2 + k_y^2}} \begin{pmatrix} k_y \\ -k_x \\ 0 \end{pmatrix} & \text{for } k_x^2 + k_y^2 > 0, \\[2em] \dfrac{\omega\mu}{2\pi} \begin{pmatrix} 0 \\ -1 \\ 0 \end{pmatrix} & \text{for } k_x^2 + k_y^2 = 0, \end{cases} \tag{B.19}$$

$$\mathbf{E}^p(\mathbf{k}) = \begin{cases} \dfrac{k_z}{2\pi\sqrt{k_x^2 + k_y^2}} \sqrt{\dfrac{\mu}{\epsilon}} \begin{pmatrix} k_x \\ k_y \\ -(k_x^2 + k_y^2)/k_z \end{pmatrix} & \text{for } k_x^2 + k_y^2 > 0, \\[2em] \dfrac{k}{2\pi}\sqrt{\dfrac{\mu}{\epsilon}} \begin{pmatrix} 1 \\ 0 \\ 0 \end{pmatrix} & \text{for } k_x^2 + k_y^2 = 0. \end{cases} \tag{B.20}$$

and

$$\mathbf{H}^s(\mathbf{k}) = \begin{cases} \dfrac{k_z}{2\pi\sqrt{k_x^2+k_y^2}} \begin{pmatrix} k_x \\ k_y \\ -(k_x^2+k_y^2)/k_z \end{pmatrix} & \text{for } k_x^2+k_y^2 > 0, \\ \dfrac{k}{2\pi} \begin{pmatrix} 1 \\ 0 \\ 0 \end{pmatrix} & \text{for } k_x^2+k_y^2 = 0, \end{cases} \quad \text{(B.21)}$$

$$\mathbf{H}^p(\mathbf{k}) = \begin{cases} \dfrac{k}{2\pi\sqrt{k_x^2+k_y^2}} \begin{pmatrix} -k_y \\ k_x \\ 0 \end{pmatrix} & \text{for } k_x^2+k_y^2 > 0, \\ \dfrac{k}{2\pi} \begin{pmatrix} 1 \\ 0 \\ 0 \end{pmatrix} & \text{for } k_x^2+k_y^2 = 0, \end{cases} \quad \text{(B.22)}$$

as follows from the curl equations (Eqs. (4.3) and (4.4)). Although arbitrary the choice of normalisation has been chosen to follow [2]. As such s and p polarised plane waves satisfy the orthogonality relations[3]

$$\langle \mathbf{E}^\nu(\mathbf{k},\mathbf{r}), \mathbf{E}^{\nu'}(\mathbf{k}',\mathbf{r}) \rangle = \left[\left(\frac{\omega\mu}{2\pi}\right)^2 \delta_{\nu,\nu'}\delta_{\nu,s} + \frac{\mu}{\epsilon}\left(\frac{|k_z|}{2\pi}\right)^2 \delta_{\nu,\nu'}\delta_{\nu,p} \right] \delta(\mathbf{k}-\mathbf{k}'), \quad \text{(B.24a)}$$

$$\langle \mathbf{H}^\nu(\mathbf{k},\mathbf{r}), \mathbf{H}^{\nu'}(\mathbf{k}',\mathbf{r}) \rangle = \left[\left(\frac{|k_z|}{2\pi}\right)^2 \delta_{\nu,\nu'}\delta_{\nu,s} + \left(\frac{k}{2\pi}\right)^2 \delta_{\nu,\nu'}\delta_{\nu,p} \right] \delta(\mathbf{k}-\mathbf{k}'). \quad \text{(B.24b)}$$

---

[3] The inner product of two vector fields is defined as

$$\langle \mathbf{U}_1(x,y), \mathbf{U}_2(x,y) \rangle = \int_{-\infty}^{\infty} \int_{-\infty}^{\infty} \mathbf{U}_1^\dagger(x,y) \cdot \mathbf{U}_2(x,y) dx dy. \quad \text{(B.23)}$$

Appendix B: Special Functions

B.3 Rectangular Waveguide Modes

Since the MODS data pits discussed in Chap. 7 are modelled as perfectly conducting, terminated rectangular waveguides in numerical simulations, a full derivation of the associated waveguide modes is presented in this section. This will involve application of the appropriate boundary conditions (introduced in Sect. B.3.1) and Maxwell's equations (performed in Sects. B.3.2 and B.3.3 for transverse electric and transverse magnetic field configurations).

B.3.1 Boundary Conditions

In general the presence of a boundary between two different media gives rise to a discontinuity in an electric or magnetic field incident upon it. Maxwell's equations can however be used to determine the difference in the fields between the two media. This is given by

$$\begin{aligned} \epsilon_1 E_1^\perp - \epsilon_2 E_2^\perp = \sigma_f, & \quad E_1^\| - E_2^\| = 0, \\ \mu_1 H_1^\perp - \mu_2 H_2^\perp = 0, & \quad H_1^\| - H_2^\| = \Lambda_f, \end{aligned} \tag{B.25}$$

where σ_f and Λ_f are the free surface charge and current density respectively and $U_k^\perp$ ($U_k^\|$) denote the component(s) of the field in the kth medium that lie perpendicular (parallel) to the boundary (see Fig. B.3a). If the second medium is conducting $\sigma_f \neq 0$ and $\Lambda_f \neq 0$ due to the free electrons in the conductor, however the field amplitudes decay exponentially via the skin effect. For perfect conductors the skin depth is zero, that is to say there exists no internal fields, i.e. $\mathbf{E}_2 = \mathbf{H}_2 = \mathbf{0}$. Under such conditions the electromagnetic boundary conditions simplify to

$$\begin{aligned} \epsilon_1 E_1^\perp = \sigma_f, & \quad E_1^\| = 0, \\ \mu_1 H_1^\perp = 0, & \quad H_1^\| = \Lambda_f. \end{aligned} \tag{B.26}$$

These boundary conditions are now applied to a rectangular waveguide configuration, as shown in Fig. B.3b in the x (y) direction and is centered on the optical axis. The permittivity distribution is thus

$$\epsilon(x, y, z) = \begin{cases} \epsilon_1 & \text{for } |x| \leq a, |y| \leq b \\ \epsilon_2 & \text{otherwise} \end{cases}. \tag{B.27}$$

It is further assumed that both media are non-magnetic such that $\mu_1 = \mu_2 = \mu_0$. When applied to the boundaries in between the waveguide core and cladding, Eq. (B.26) become

$$\left. \begin{aligned} \epsilon_1 E_{x1} = \sigma_f, & \quad E_{y1} = 0, & \quad E_{z1} = 0 \\ H_{x1} = 0, & \quad H_{y1} = \Lambda_f, & \quad H_{z1} = \Lambda_f \end{aligned} \right\} \text{ for } |x| = a, |y| \leq b, \tag{B.28}$$

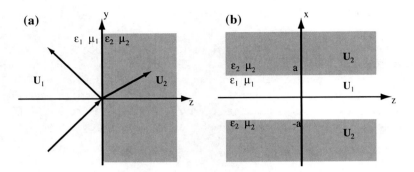

Fig. B.3 **a** Electromagnetic field incident upon a conducting medium. **b** Cross section through a *rectangular* waveguide

and

$$\left.\begin{array}{ll} E_{x1} = 0, & \epsilon_1 E_{y1} = \sigma_f, \quad E_{z1} = 0 \\ H_{x1} = \Lambda_f, & H_{y1} = 0, \quad H_{z1} = \Lambda_f \end{array}\right\} \text{ for } |x| \le a, |y| = b. \quad \text{(B.29)}$$

Writing out Eqs. (4.3) and (4.4) component-wise in full gives

$$\frac{\partial E_z}{\partial y} - \frac{\partial E_y}{\partial z} = i\mu\omega H_x, \qquad \frac{\partial H_z}{\partial y} - \frac{\partial H_y}{\partial z} = -i\epsilon\omega E_x, \quad \text{(B.30a)}$$

$$-\frac{\partial E_z}{\partial x} + \frac{\partial E_x}{\partial z} = i\mu\omega H_y, \qquad -\frac{\partial H_z}{\partial x} + \frac{\partial H_x}{\partial z} = -i\epsilon\omega E_y, \quad \text{(B.30b)}$$

$$\frac{\partial E_y}{\partial x} - \frac{\partial E_x}{\partial y} = i\mu\omega H_z, \qquad \frac{\partial H_y}{\partial x} - \frac{\partial H_x}{\partial y} = -i\epsilon\omega E_z, \quad \text{(B.30c)}$$

which upon letting $U_j = U_j^+ + U_j^-$ where $U_j^\pm(x, y, z) = X_j(x) Y_j(y) C_j^\pm \exp(\pm i\kappa_z z)$ and various algebraic manipulations yields

$$E_x = \frac{i}{k^2 - \kappa_z^2} \left[\mu\omega \frac{\partial}{\partial y} (H_z^+ + H_z^-) + \kappa_z \frac{\partial}{\partial x} (E_z^+ - E_z^-) \right], \quad \text{(B.31a)}$$

$$E_y = \frac{-i}{k^2 - \kappa_z^2} \left[\mu\omega \frac{\partial}{\partial x} (H_z^+ + H_z^-) - \kappa_z \frac{\partial}{\partial y} (E_z^+ - E_z^-) \right], \quad \text{(B.31b)}$$

$$H_x = \frac{-i}{k^2 - \kappa_z^2} \left[\epsilon\omega \frac{\partial}{\partial y} (E_z^+ + E_z^-) - \kappa_z \frac{\partial}{\partial x} (H_z^+ - H_z^-) \right], \quad \text{(B.31c)}$$

$$H_y = \frac{i}{k^2 - \kappa_z^2} \left[\epsilon\omega \frac{\partial}{\partial x} (E_z^+ + E_z^-) + \kappa_z \frac{\partial}{\partial y} (H_z^+ - H_z^-) \right]. \quad \text{(B.31d)}$$

It should be noted that so as to distinguish the waveguide mode propagation coefficients from those of plane waves, the notation $\boldsymbol{\kappa} = (\kappa_x, \kappa_y, \kappa_z)$ is now used, where $\boldsymbol{\kappa} \cdot \boldsymbol{\kappa}^* = k^2$.

The importance of these equations lies in the fact that the field in a waveguide can be decomposed into a summation of transverse magnetic (TM) and transverse electric (TE) modes [5] defined respectively as having zero magnetic or electric field in the propagation direction (taken as per normal convention as the positive z direction). For TM modes, for which $H_z^+ = H_z^- = 0$, Eqs. (B.31) become

$$E_x^{TM} = \frac{i\kappa_z}{k^2 - \kappa_z^2} \frac{\partial}{\partial x} (E_z^{TM+} - E_z^{TM-}), \tag{B.32a}$$

$$E_y^{TM} = \frac{i\kappa_z}{k^2 - \kappa_z^2} \frac{\partial}{\partial y} (E_z^{TM+} - E_z^{TM-}), \tag{B.32b}$$

$$H_x^{TM} = \frac{i\epsilon\omega}{k^2 - \kappa_z^2} \frac{\partial}{\partial y} (E_z^{TM+} + E_z^{TM-}), \tag{B.32c}$$

$$H_y^{TM} = \frac{-i\epsilon\omega}{k^2 - \kappa_z^2} \frac{\partial}{\partial x} (E_z^{TM+} + E_z^{TM-}), \tag{B.32d}$$

whilst for TE modes ($E_z^+ = E_z^- = 0$) they become

$$E_x^{TE} = \frac{-i\mu\omega}{k^2 - \kappa_z^2} \frac{\partial}{\partial y} (H_z^{TE+} + H_z^{TE-}), \tag{B.33a}$$

$$E_y^{TE} = \frac{i\mu\omega}{k^2 - \kappa_z^2} \frac{\partial}{\partial x} (H_z^{TE+} + H_z^{TE-}), \tag{B.33b}$$

$$H_x^{TE} = \frac{i\kappa_z}{k^2 - \kappa_z^2} \frac{\partial}{\partial x} (H_z^{TE+} - H_z^{TE-}), \tag{B.33c}$$

$$H_y^{TE} = \frac{i\kappa_z}{k^2 - \kappa_z^2} \frac{\partial}{\partial y} (H_z^{TE+} - H_z^{TE-}). \tag{B.33d}$$

Via Eqs. (B.32) and (B.33) it is hence possible to relate the transverse components of a field to the longitudinal field component. Derivation of the waveguide modes thus reduces to a derivation of the supported forms of the longitudinal electric and magnetic field components.

B.3.2 Transverse Magnetic Modes

Consider first TM modes for which $H_{z1} = 0$ for all (x, y, z). When the boundary condition $E_{z1}(\pm a, y) = 0$ and Eqs. (B.15) are combined, the equation

$$0 = \left(A_z^+ e^{\pm i\kappa_x a} + A_z^- e^{\mp i\kappa_x a}\right)\left(B_z^+ e^{i\kappa_y y} + B_z^- e^{i\kappa_y y}\right)\left(C_z^+ e^{i\kappa_z z} + C_z^- e^{-i\kappa_z z}\right),$$
$$0 = \left(A_z^+ e^{\pm i\kappa_x a} + A_z^- e^{\mp i\kappa_x a}\right),$$

follows and hence

$$A_z^+ = -A_z^- \exp(2i\kappa_x a), \tag{B.34a}$$

$$A_z^+ = -A_z^- \exp(-2i\kappa_x a). \tag{B.34b}$$

Dividing these equations yields $\exp(4i\kappa_x a) = 1$ such that

$$\kappa_x = \frac{m\pi}{2a} \quad m \in \mathbb{Z}, \tag{B.35}$$

which upon substitution subsequently gives $A_z^+ = (-1)^{m+1} A_z^-$. Letting $A_z^+ = A = 1$ without loss of generality, these results can be combined to give

$$X_z(x) = \begin{cases} 2Ai \sin(\kappa_x x) & \text{for } m \text{ even} \\ 2A \cos(\kappa_x x) & \text{for } m \text{ odd.} \end{cases} \tag{B.36}$$

which can be condensed into the single expression

$$X_{zm}^{\text{TM}}(x) = [1 + (-1)^m] i \sin(\kappa_x x) + [1 - (-1)^m] \cos(\kappa_x x). \tag{B.37}$$

Similarly, applying the boundary condition $E_{z1}(x, \pm b) = 0$ gives

$$Y_{zn}^{\text{TM}}(y) = [1 + (-1)^n] i \sin(\kappa_y y) + [1 - (-1)^n] \cos(\kappa_y y), \tag{B.38}$$

where $\kappa_y = \frac{n\pi}{2b}$ for $n \in \mathbb{Z}$. The longitudinal field component for a single TM mode can thus be written

$$\begin{aligned} E_{zmn}^{\text{TM}}(x, y, z) &= \{[1 + (-1)^m] i \sin(\kappa_x x) + [1 - (-1)^m] \cos(\kappa_x x)\} \\ &\times \{[1 + (-1)^n] i \sin(\kappa_y y) + [1 - (-1)^n] \cos(\kappa_y y)\} \\ &\times \{C_z^{\text{TM}+} \exp(i\kappa_z z) + C_z^{\text{TM}-} \exp(-i\kappa_z z)\}, \end{aligned} \tag{B.39}$$

where mn denotes the mode indices. From Eqs. (B.32) the remaining field components follow:

$$E_{xmn}^{\text{TM}} = \frac{i\kappa_z}{k^2 - \kappa_z^2} (C_z^{\text{TM}+} e^{i\kappa_z z} - C_z^{\text{TM}-} e^{-i\kappa_z z}) \frac{\partial}{\partial x} X_{zm}^{\text{TM}}(x) Y_{zn}^{\text{TM}}(y), \tag{B.40a}$$

$$E_{ymn}^{\text{TM}} = \frac{i\kappa_z}{k^2 - \kappa_z^2} (C_z^{\text{TM}+} e^{i\kappa_z z} - C_z^{\text{TM}-} e^{-i\kappa_z z}) X_{zm}^{\text{TM}}(x) \frac{\partial}{\partial y} Y_{zn}^{\text{TM}}(y), \tag{B.40b}$$

Appendix B: Special Functions

$$H_{xmn}^{TM} = \frac{-i\epsilon\omega}{k^2 - \kappa_z^2} (C_z^{TM+} e^{i\kappa_z z} + C_z^{TM-} e^{-i\kappa_z z}) X_{zm}^{TM}(x) \frac{\partial}{\partial y} Y_{zn}^{TM}(y), \tag{B.40c}$$

$$H_{ymn}^{TM} = \frac{i\epsilon\omega}{k^2 - \kappa_z^2} (C_z^{TM+} e^{i\kappa_z z} + C_z^{TM-} e^{-i\kappa_z z}) \frac{\partial}{\partial x} X_{zm}^{TM}(x) Y_{zn}^{TM}(y), \tag{B.40d}$$

where

$$\frac{\partial}{\partial x} X_{zm}^{TM}(x) = [1 + (-1)^m] \kappa_x i \cos(\kappa_x x) - [1 - (-1)^m] \kappa_x \sin(\kappa_x x), \tag{B.41a}$$

$$\frac{\partial}{\partial y} Y_{zn}^{TM}(y) = [1 + (-1)^n] \kappa_y i \cos(\kappa_y y) - [1 - (-1)^n] \kappa_y \sin(\kappa_y y). \tag{B.41b}$$

Since if $m = 0$ or $n = 0$ then $E_x^{TM} = E_y^{TM} = H_x^{TM} = H_y^{TM} = 0$, the lowest TM mode is the $(m, n) = (1, 1)$ mode denoted TM_{11}.

B.3.3 Transverse Electric Modes

Similar analysis can be performed for TE modes by applying the boundary conditions $H_{x1}(\pm a, y) = 0$ and $H_{y1}(x, \pm b) = 0$ in turn. From Eqs. (B.33) these boundary conditions imply

$$\frac{\partial H_z}{\partial x} = 0 \quad \text{for } |x| = a, |y| \le b, \tag{B.42}$$

and

$$\frac{\partial H_z}{\partial y} = 0 \quad \text{for } |x| \le a, |y| = b. \tag{B.43}$$

Following analogous manipulations as given in Sect. B.3.2 the longitudinal magnetic field component is given by

$$H_{zmn}^{TE}(x, y, z) = \{[1 - (-1)^m] i \sin(\kappa_x x) + [1 + (-1)^m] \cos(\kappa_x x)\}$$
$$\{[1 - (-1)^n] B^{TE} i \sin(\kappa_y y) + [1 + (-1)^n] \cos(\kappa_y y)\}$$
$$\times \{C_z^{TE+} \exp(i\kappa_z z) + C_z^{TE-} \exp(-i\kappa_z z)\}. \tag{B.44}$$

From Eqs. (B.33) the transverse field components are found to be

$$E_{xmn}^{TE} = \frac{i\mu\omega}{k^2 - \kappa_z^2} (C_z^{TE+} e^{i\kappa_z z} + C_z^{TE-} e^{-i\kappa_z z}) X_{zm}^{TE}(x) \frac{\partial}{\partial y} Y_{zn}^{TE}(y), \tag{B.45a}$$

$$E_{ymn}^{TE} = \frac{-i\mu\omega}{k^2 - \kappa_z^2} (C_z^{TE+} e^{i\kappa_z z} + C_z^{TE-} e^{-i\kappa_z z}) \frac{\partial}{\partial x} X_{zm}^{TE}(x) Y_{zn}^{TE}(y), \tag{B.45b}$$

$$H_{xmn}^{TE} = \frac{i\kappa_z}{k^2 - \kappa_z^2}(C_z^{TE+}e^{i\kappa_z z} - C_z^{TE-}e^{-i\kappa_z z})\frac{\partial}{\partial x}X_{zm}^{TE}(x)Y_{zn}^{TE}(y), \tag{B.45c}$$

$$H_{ymn}^{TE} = \frac{i\kappa_z}{k^2 - \kappa_z^2}(C_z^{TE+}e^{i\kappa_z z} - C_z^{TE-}e^{-i\kappa_z z})X_{zm}^{TE}(x)\frac{\partial}{\partial y}Y_{zn}^{TE}(y), \tag{B.45d}$$

where

$$X_{zm}^{TE}(x) = [1 - (-1)^m]i\sin(\kappa_x x) + [1 + (-1)^m]\cos(\kappa_x x), \tag{B.46a}$$

$$Y_{zn}^{TE}(y) = [1 - (-1)^n]i\sin(\kappa_y y) + [1 + (-1)^n]\cos(\kappa_y y), \tag{B.46b}$$

$$\frac{\partial}{\partial x}X_{zm}^{TE}(x) = [1 - (-1)^m]\kappa_x i\cos(\kappa_x x) - [1 + (-1)^m]\kappa_x \sin(\kappa_x x), \tag{B.46c}$$

$$\frac{\partial}{\partial y}Y_{zn}^{TE}(y) = [1 - (-1)^n]\kappa_y i\cos(\kappa_y y) - [1 + (-1)^n]\kappa_y \sin(\kappa_y y). \tag{B.46d}$$

To conclude this section the relationship between the constants $C_z^{TE\pm}$ and $C_z^{TM\pm}$ that exists for a terminated waveguide is derived. It is assumed that the rectangular waveguide is terminated by a perfect conductor at a depth D from the waveguide entrance interface. Exploiting the resulting boundary condition $\mathbf{E}_\parallel(x, y, D) = \mathbf{0}$, which must hold for each waveguide mode separately, and by evaluating Eqs. (B.40) and (B.45) at the terminating surface ($z = D$) yields

$$0 = \frac{i\kappa_z}{k^2 - \kappa_z^2}(C_z^{TM+}e^{i\kappa_z D} - C_z^{TM-}e^{-i\kappa_z D})\frac{\partial}{\partial x}X_{zm}^{TM}(x)Y_{zn}^{TM}(y), \tag{B.47a}$$

$$0 = \frac{i\kappa_z}{k^2 - \kappa_z^2}(C_z^{TM+}e^{i\kappa_z D} - C_z^{TM-}e^{-i\kappa_z D})X_{zm}^{TM}(x)\frac{\partial}{\partial y}Y_{zn}^{TM}(y), \tag{B.47b}$$

$$0 = \frac{i\mu\omega}{k^2 - \kappa_z^2}(C_z^{TE+}e^{i\kappa_z D} + C_z^{TE-}e^{-i\kappa_z D})X_{zm}^{TE}(x)\frac{\partial}{\partial y}Y_{zn}^{TE}(y), \tag{B.47c}$$

$$0 = \frac{-i\mu\omega}{k^2 - \kappa_z^2}(C_z^{TE+}e^{i\kappa_z D} + C_z^{TE-}e^{-i\kappa_z D})\frac{\partial}{\partial y}X_{zm}^{TE}(x)Y_{zn}^{TE}(y), \tag{B.47d}$$

which reduce to

$$C_z^{TM-} = C_z^{TM+}e^{2i\kappa_z D}, \tag{B.48a}$$

$$C_z^{TE-} = -C_z^{TE+}e^{2i\kappa_z D}. \tag{B.48b}$$

Upon normalising such that $C_z^{\mu+} = 1$ ($\mu =$ TE or TM), the $Z_z^\mu(z)$ terms in the original separable solution to the wave equation (c.f. Eq. (B.15c)) become

$$Z_z^{TM}(z) = 2\cos[\kappa_z(z - D)], \tag{B.49a}$$

$$Z_z^{TE}(z) = 2\sin[\kappa_z(z - D)], \tag{B.49b}$$

where a global phase term $\exp(i\kappa_z D)$ has been dropped due to its physical insignificance.

Appendix B: Special Functions

B.3.4 Orthogonality of Waveguide Modes

In a similar fashion to plane waves, it can be shown that the transverse field components of the waveguide modes obey a number of orthogonality relations. To determine the normalisation constants required for solution of the set of linear equations described by Eq. (7.18), the overlap integrals, or inner products, between modes must be calculated. Although requiring tedious manipulations, analytic answers can be found and are given by

$$\langle \mathbf{E}_{\|mn}^{TM}, \mathbf{E}_{\|pq}^{TM} \rangle = \left| \frac{2\pi\kappa_z}{k^2 - \kappa_z^2} \right|^2 \sin^2 \kappa_z D \left[m^2 \frac{b}{a} + n^2 \frac{a}{b} \right] \delta_{mp} \delta_{nq}, \tag{B.50a}$$

$$\langle \mathbf{E}_{\|mn}^{TE}, \mathbf{E}_{\|pq}^{TE} \rangle = \left| \frac{2\pi\omega\mu}{k^2 - \kappa_z^2} \right|^2 \sin^2 \kappa_z D \left[m^2 \frac{b}{a} + n^2 \frac{a}{b} \right] \delta_{mp} \delta_{nq}, \tag{B.50b}$$

$$\langle \mathbf{H}_{\|mn}^{TM}, \mathbf{H}_{\|pq}^{TM} \rangle = \left| \frac{2\pi\epsilon\omega}{k^2 - \kappa_z^2} \right|^2 \cos^2 \kappa_z D \left[m^2 \frac{b}{a} + n^2 \frac{a}{b} \right] \delta_{mp} \delta_{nq}, \tag{B.50c}$$

$$\langle \mathbf{H}_{\|mn}^{TE}, \mathbf{H}_{\|pq}^{TE} \rangle = \left| \frac{2\pi\kappa_z}{k^2 - \kappa_z^2} \right|^2 \cos^2 \kappa_z D \left[m^2 \frac{b}{a} + n^2 \frac{a}{b} \right] \delta_{mp} \delta_{nq}. \tag{B.50d}$$

All other overlap integrals are identically zero.

B.3.5 Plane Wave and Waveguide Mode Coupling

Further to the orthogonality relations expressed in Eqs. (B.24) and (B.50), the coupling strength between the transverse components of plane wave modes and waveguide modes is required for solution of Eq. (7.18). Analytic expressions are again derivable and are given by

$$\langle \mathbf{E}_{\|}^{s}(\mathbf{k}), \mathbf{E}_{\|mn}^{TM} \rangle = \frac{i\kappa_z\omega\mu}{2\pi\sqrt{k_x^2 + k_y^2}} \frac{\sin \kappa_z D}{k^2 - \kappa_z^2} \left[k_y\kappa_x\chi_{m,x}^{TM-}\chi_{n,y}^{TM+} - k_x\kappa_y\chi_{m,x}^{TM+}\chi_{n,y}^{TM-} \right],$$

$$\langle \mathbf{E}_{\|}^{s}(\mathbf{k}), \mathbf{E}_{\|mn}^{TE} \rangle = \frac{i\omega^2\mu^2}{2\pi\sqrt{k_x^2 + k_y^2}} \frac{\sin \kappa_z D}{k^2 - \kappa_z^2} \left[k_x\kappa_x\chi_{m,x}^{TE-}\chi_{n,y}^{TE+} + k_y\kappa_y\chi_{m,x}^{TE+}\chi_{n,y}^{TE-} \right],$$

$$\langle \mathbf{E}_\|^p(\mathbf{k}), \mathbf{E}_{\|mn}^{TM}\rangle = \frac{ik_z\kappa_z}{2\pi\sqrt{k_x^2+k_y^2}} \sqrt{\frac{\mu}{\epsilon}} \frac{\sin\kappa_z D}{k^2-\kappa_z^2} \left[k_x\kappa_x\chi_{m,x}^{TM-}\chi_{n,y}^{TM+} + k_y\kappa_y\chi_{m,x}^{TM+}\chi_{n,y}^{TM-} \right],$$

$$\langle \mathbf{E}_\|^p(\mathbf{k}), \mathbf{E}_{\|mn}^{TE}\rangle = \frac{ik_z\omega\mu}{2\pi\sqrt{k_x^2+k_y^2}} \sqrt{\frac{\mu}{\epsilon}} \frac{\sin\kappa_z D}{k^2-\kappa_z^2} \left[k_x\kappa_y\chi_{m,x}^{TE+}\chi_{n,y}^{TE-} - k_y\kappa_x\chi_{m,x}^{TE+-}\chi_{n,y}^{TE+} \right],$$

$$\langle \mathbf{H}_\|^s(\mathbf{k}), \mathbf{H}_{\|mn}^{TM}\rangle = \frac{ik_z\omega\epsilon}{2\pi\sqrt{k_x^2+k_y^2}} \frac{\cos\kappa_z D}{k^2-\kappa_z^2} \left[k_y\kappa_x\chi_{m,x}^{TM-}\chi_{n,y}^{TM+} - k_x\kappa_y\chi_{m,x}^{TM+}\chi_{n,y}^{TM-} \right],$$

$$\langle \mathbf{H}_\|^s(\mathbf{k}), \mathbf{H}_{\|mn}^{TE}\rangle = \frac{ik_z\kappa_z}{2\pi\sqrt{k_x^2+k_y^2}} \frac{\cos\kappa_z D}{k^2-\kappa_z^2} \left[k_x\kappa_x\chi_{m,x}^{TE-}\chi_{n,y}^{TE+} + k_y\kappa_y\chi_{m,x}^{TE+}\chi_{n,y}^{TE-} \right],$$

$$\langle \mathbf{H}_\|^p(\mathbf{k}), \mathbf{H}_{\|mn}^{TM}\rangle = \frac{ik\omega\epsilon}{2\pi\sqrt{k_x^2+k_y^2}} \frac{\cos\kappa_z D}{k^2-\kappa_z^2} \left[k_x\kappa_x\chi_{m,x}^{TM-}\chi_{n,y}^{TM+} + k_y\kappa_y\chi_{m,x}^{TM+}\chi_{n,y}^{TM-} \right],$$

$$\langle \mathbf{H}_\|^p(\mathbf{k}), \mathbf{H}_{\|mn}^{TE}\rangle = \frac{ik\kappa_z}{2\pi\sqrt{k_x^2+k_y^2}} \frac{\cos\kappa_z D}{k^2-\kappa_z^2} \left[k_x\kappa_y\chi_{m,x}^{TE+}\chi_{n,y}^{TE-} - k_y\kappa_x\chi_{m,x}^{TE+-}\chi_{n,y}^{TE+} \right],$$

where

$$\chi_{p,j}^{TM+} = i[1+(-1)^p]s_p^j + [1-(-1)^p]c_p^j, \tag{B.51a}$$

$$\chi_{p,j}^{TM-} = i[1+(-1)^p]c_p^j - [1-(-1)^p]s_p^j, \tag{B.51b}$$

$$\chi_{p,j}^{TE+} = i[1-(-1)^p]s_p^j + [1+(-1)^p]c_p^j, \tag{B.51c}$$

$$\chi_{p,j}^{TE-} = i[1-(-1)^p]c_p^j - [1+(-1)^p]s_p^j, \tag{B.51d}$$

and

$$c_p^j = \int_{-L}^{L} \cos(\kappa_j t) \exp[ik_j t] dt,$$

$$= \begin{cases} i^p[1+(-1)^p]\dfrac{k_j}{k_j^2-\kappa_j^2}\sin k_j L \\ \quad +i^p[1-(-1)^p]\dfrac{i\kappa_j}{k_j^2-\kappa_j^2}\cos k_j L, & k_j \neq \kappa_j \\ L, & k_j = \pm\kappa_j \neq 0 \\ 2L, & k_j = \kappa_j = 0 \end{cases} \tag{B.52}$$

$$s_p^j = \int_{-L}^{L} \sin(\kappa_j t) \exp[ik_j t] dt,$$

$$= \begin{cases} -i^p[1+(-1)^p]\dfrac{i\kappa_j}{k_j^2-\kappa_j^2}\sin k_j L \\ +i^p[1-(-1)^p]\dfrac{k_j}{k_j^2-\kappa_j^2}\cos k_j L, & k_j \neq \kappa_j \\ \mp iL, & k_j = \pm\kappa_j \neq 0 \\ 2L, & k_j = \kappa_j = 0 \end{cases} \quad (B.53)$$

where $L = a$ (or b) for $j = x$ (or y), respectively.

References

1. M.L. Boas, Mathematical Methods in the Physical Sciences, 2nd edn. (Wiley, New York, 1983)
2. J.M. Brok, An analytic approach to electromagnetic scattering problems. Ph.D. thesis, 2007
3. B.R. Frieden (1971) Evaluation, design and extrapolation methods for optical signals, based on the prolate functions, Progress in Optics IX. North-Holland Publishing Co., Amsterdam
4. J.C. Heurtley, in *Proceedings of Symposium on Quasi-Optics*, ed. by J. Fox. Hyperspheroidal Functions-Optical Resonators with Circular Mirrors, (Polytechnic Press, 1964) pp. 367–371
5. J.D. Jackson, *Classical Electrodynamics*. (Wiley, New York, 1998)
6. H.J. Landau, H.O.Pollak, Prolate spheroidal wave functions, Fourier analysis and uncertainty II. Bell Syst. Tech. J. **40**, 65–84 (1961)
7. H.J. Landau, H.O. Pollak, Prolate spheroidal wave functions, Fourier analysis and uncertainty III The dimension of the space of essentially time- and band-limited signals. Bell Syst. Tech. J. **41**, 1295–1336 (1962)
8. A.W. Lohmann, R.G. Dorsch, D. Mendlovic, Z. Zalevsky, C. Ferreira, Space-bandwidth product of optical signals and systems. J. Opt. Soc. Am. A **13**, 470–473, (1996)
9. M.A. Neifeld, Information, resolution, and space bandwidth product. Opt. Lett. **18**, 1477–1479 (1998)
10. D. Slepian, Prolate spheroidal wave functions, Fourier analysis and uncertainty IV Extensions to many dimensions; generalised prolate spheroidal functions. Bell System Tech. J. **43**, 3009–3057 (1964)
11. D. Slepian, Some comments on Fourier analysis, uncertainty and modeling. SIAM Review **25**, 379–393 (2000)
12. D. Slepian, H.O. Pollak, Prolate spheroidal wave functions, Fourier analysis and uncertainty I. Bell System. Tech. J. **40**, 43–64 (1961)